Raquel Aparecida Mendes Lima
Adriana Malvasio

Effects of pesticides on tropical amphibians

Raquel Aparecida Mendes Lima
Adriana Malvasio

Effects of pesticides on tropical amphibians

Unraveling environmental issues

ScienciaScripts

Imprint

Cover image: www.ingimage.com

This book is a translation from the original published under ISBN 978-613-9-65434-5.

Publisher:
Sciencia Scripts
is a trademark of
Dodo Books Indian Ocean Ltd. and OmniScriptum S.R.L publishing group

120 High Road, East Finchley, London, N2 9ED, United Kingdom
Str. Armeneasca 28/1, office 1, Chisinau MD-2012, Republic of Moldova, Europe
Printed at: see last page
ISBN: 978-620-7-77663-4

SUMMARY

AUTHORS

Raquel Aparecida Mendes Lima (Main Author)

Graduated in Environmental Engineering (2004), Specialist in Water Resources (2006), Master in Environmental Sciences (2010) from UFT and PhD student in ecotoxicology at UA/PT. Professor of the Environment at IFTO. She has experience in managing environmental policies and standards; preparing, executing and managing projects in the areas of environmental planning and management, water resources, environmental education, agribusiness; socio-environmental perception; ecotoxicology and functional biodiversity, with an emphasis on amphibians; environmental impact assessment. She worked as an environmental engineer for the Tocantins State Government, coordinating socio-environmental impact studies; ecotourism; climate change. She has conducted participatory workshops and negotiated public environmental projects with the World Bank.

Adriana Malvasio (Coauthor)

He holds a bachelor's degree in Biological Sciences from Mackenzie Presbyterian University (MACK), a master's degree in Biological Sciences (Zoology) from the Zoology Department of the Biosciences Institute of the University of Sao Paulo (IB/USP) and a doctorate in Biological Sciences (Zoology) from the Zoology Department of the Biosciences Institute of the University of Sao Paulo (IB/USP). She did post-doctoral work at the Animal Ecology Laboratory of the Biology Department of the Luiz de Queiróz College of Agriculture (Esalq/USP). She is currently an associate professor at the Federal University of Tocantins (UFT), working on the undergraduate courses in Environmental Engineering and the postgraduate programs in Environmental Sciences and Ecology of Ecotones. He has experience in Zoology and Animal Ecology, with an emphasis on Chelonians and Crocodilians.

"We learn knowledge from teachers and books.

wisdom is learned from life and from the humble"!

Cora Coralina.

I dedicate

To my parents (João Ferreira Lima and Nilda Mendes Lima) and my daughter Stefanie, for their unconditional love...

Collaborators:

1. Erton Moraes
2. Eugênio Hengdes Guerreiro
3. Paulo Henrique Alves Rodrigues
4. André Costa Pereira
5. Josefa Nascimento
6. Elineide Marques
7. Valéria Gomes Momenté
8. Croque (North Region Crocodile and Chelonian Research Group)/UFT

THANKS

- To God, for my existence...
- To Professor Dr. Carla Cassini - Federal University of Espirito Santo, for providing material that contributed to this research;
- The Federal University of Tocantins, colleagues, teachers and staff of the master's course in environmental sciences;
- The University of Portugal - UA and Professor Amadeu Soares for their support during the exchange program;
- To the Department of Water Resources and the Environment, for their support during my master's degree, especially to Mr. Belizàrio Franco Neto, Mr. Anizio Pedreira and Mr. Anilton;
- To Professor Paula Benevides, Professor Rafael and Deusiano and the other technicians of the Microbiology laboratory at the Federal University of Tocantins;
- To the technician Antônio Souza da Silva, from the Histopathology laboratory at the Federal University of Goiás, for his help in preparing the histological slides;
- To dear Ruberval Barbosa, Agricultural Engineer, for his experience and important tips that contributed to this research.
- To my friends, for their support, affection, dedication and relaxation, especially during the most difficult times (Geise Quirino, Ana Carla, Fabiane Ribeiro, Clairton Thomazi, Ediclea Lopes de Araùjo, Cristiane Peres and Weilan Cris Brito, Sebastiao Bastos, Màrio Pombal, Eduardo Ferreira and Rita from UA/PT).

PRESENTATION

In the state of Tocantins, the use of agricultural products in crops is common, which can be seen in the municipality of Lagoa da Confusao, known for the number of agricultural projects developed in the region. The aim of this study is to assess the contamination that can be caused to *Leptodactylus latrans* by the presence of pesticides in the Urubu river sub-basin - TO, with the specific objectives of carrying out a quantitative and qualitative survey of producers and pesticide stores that were approached; as well as to carry out a histological analysis of the integument, liver and reproductive system of *Leptodactylus latrans,* caught at three study sites, Cristalândia and Lagoa da Confusao - Tocantins. The Urubu River is part of the Formoso River sub-basin, is about 2,702 km long^2 and its main course is 130 km to its mouth. The study is located between the geographic coordinates W 49° 41' 17.54"; S 10° 50' 49.32" and W 49° 2' 43.99"; S 10° 38' 13.84". This study was authorized by the Ethics Committee of the Lutheran University of Brazil, as well as by the Center for the Conservation and Management of Reptiles and Amphibians - RAN/ICMBio. Sampling took place during 2009, with nocturnal visits near the Urubu River. Captures were made manually with the aid of a spear. This procedure was carried out over two or three consecutive nights. Of the individuals captured, 5 (five) were apparently healthy and 100% with malformations, which were sent to the UFG Histological Analysis Laboratory. The histological study prioritized slides from the reproductive system, the epithelium and the liver, since according to the literature, they can be widely affected by contamination from pesticides. Semi-structured interviews were conducted in the vicinity of points 2 and 3 with farmers, and a survey was also carried out at the place where empty pesticide containers are returned. From this, it could be seen that in areas where deforestation has taken place and where flood-irrigated rice is grown, there were more *L. latrans, as* well as smaller individuals. This differs from the situation observed in the native vegetation area, where a greater number of larger individuals were captured. A similar situation occurred at point 3, with the presence of grass; the individuals found were of different sizes and in greater numbers. During the interviews, it became apparent that many of the farmers responsible for planting the crops, despite stating that it would not be possible to produce without using pesticides, were afraid to divulge the products they used. Histopathological changes were identified in the liver and skin of *L. latrans* at the three sites studied. **Key words:** Agricultural plantations. Native vegetation. Pesticides. Leptodactylidae.

CHAPTER I GENERAL ASPECTS OF PESTICIDE USE IN AGRICULTURE AND ITS EFFECTS ON AMPHIBIAN POPULATIONS

INTRODUCTION

1 - Agriculture and pesticides

Around 50% of the Brazilian population is still rural and a third still cooks on a wood-burning stove. This is seen as a loss of importance for this sector, whereas in fact it is a central axis of economic activity in the world. Today we don't just talk about food, we talk about food, animal feed, fuel and fiber, to which medicinal uses have been added more recently. This reflects the multifunctionality of agriculture and a broader understanding of its challenges (MCNEELY and SCHERR, 2009).

Some positive points in Brazilian agriculture, especially in the last decade, is the growth in grain production on small properties (less than 100 hectares), which grew by 5.8% between 1989 and 1999. On the other hand, there are also advances in agroforestry and organic farming projects, many of which involve civil society, governments and private initiative, as well as farmers (CAMARGO *et al.*, 2002).

Agroforestry systems are ways of using the land, combining trees, shrubs and palms in agricultural crops and pastures, thus favoring the sedentarization of farmers and improving their living conditions. The spread of agroforestry systems also contributes to the conservation of the remaining native forests and their biodiversity, as well as the inclusion of rural populations in the market economy. On the other hand, the tree components of agroforestry systems can be considered a mechanism for sustainable development. In the surroundings of Conservation Units, the spread of agroforestry systems encourages farmers to remain on their land and therefore tends to reduce external pressures on protected areas (DUBOIS, 2002).

Bean producers, not only in the region of Lagoa da Confusâo - TO, but throughout Brazil, still choose to plant beans without any selection criteria because of the investments made in technology. Producers in the region have not yet woken up to the potential that the region has in terms of producing bean seeds (CERRADO RURAL, 2009).

On the other hand, there is the implementation of large projects that involve the transformation of water geography in order to promote economic growth, which modify both the physical, chemical and biological characteristics of water resources, as well as affecting social relations (NAVAL *et al.*, 2008).

As for Brazilian agriculture, some data reveal its unsustainability: one percent of landowners own 45.1% of the agricultural area, while 89.3% of small landowners, who have less than 100 hectares, control only 20% of the agricultural area. Furthermore, it can be seen that the consumption of pesticides grew by more than 276% between 1960 and 1991 and the use of pesticides per planted area grew by 21.59 between 1997 and 2000. Agriculture is also responsible for the large consumption of a scarce and strategically important resource: water. Where 59% of the water produced in Brazil is destined for agriculture, while industry consumes 19% and domestic uses another 22% (CAMARGO *et al.*, 2002).

Due to the widespread use of pesticides in food production, there is an association between exposure and the development of lung, breast, testicular, thyroid, prostate, ovarian and hematopoietic cancers (PIMENTEL, 1996). In Brazil, there is a high mortality rate for cancer of the larynx, stomach, esophagus and leukemia in farmers who are exposed to the product (MEYER *et al.*, 2003).

While more than 2 billion grains are produced on the planet, there are more than 800 million people going hungry, 10 million children die from malnutrition and related diseases; large institutional investors acquire huge tracts of land that are neither cultivated nor allowed to be cultivated, and in this context, grains become the object of speculative mechanisms, with price jumps that disorganize producers and generate food catastrophes in importing countries (MCNEELY and SCHERR, 2009).

As a result of the ways in which humans use water, the biodiversity of freshwater ecosystems is more threatened than that of terrestrial ecosystems. The negative impacts of agricultural production include pollution from agrochemicals, soil and grazing land degradation and the loss of important habitats and associated wild biodiversity. Agricultural intensification in most of central, western and northern Europe is leading to declines in the wildlife that inhabits farmland (SOTHERTON, 1998).

Brazil is considered to be one of the biggest users of pesticides and has seen an increase in agricultural land over the last 30 years (SPADOTTO *et al.*, 1996).

With the Second World War, various chemical products were synthesized, and there was a great growth and development of the chemical industry. Many chemicals were used to control pests and vectors. The use of these substances intensified with the mechanization of farming and led, in addition to the rural exodus and concentration of properties, to an intense process of exposure of populations to these agents (CETESB, 2009).

Brazil imports around 60% of the fertilizers used in agriculture in the country, which may be inhibiting growth in the sector. According to a report produced by the Agriculture Commission of the Chamber of Deputies in June 2008, the Brazilian market is the fourth largest, with growth of 70% in the last ten years (TERRA VIVA, 2009).

Over time and through studies, scientists discovered that dichlorodiphenyltrichloroethane - DDT - was the most abundant pollutant in the ecosystem. It persists in the body and with each new dose it simply adds to the existing load, the same happens in biological organisms and in many unexpected organisms. Contamination after spraying with DDT reaches forests and even small creatures, but the highest levels were found in carnivorous animals in which it persists for longer than in herbivorous animals. The animals analyzed were in areas that had been sprayed for six years. DDT is carcinogenic, affects the nervous system, the liver and also has an effect on fertility (TAYLOR, 1978).

The use of herbicides, especially with the advent of no-till farming, has increased with modern agriculture. Some pesticides are so resistant to microbial attack that they can persist in the environment for more than ten years. Several species of bacteria and fungi have the ability to degrade, but the disappearance of a substance from an ecosystem does not necessarily mean that it has been degraded by microorganisms, as the loss of the pesticide can also occur through volatilization, leaching or spontaneous chemical degradation (CETESB, 2009).

Pesticides can be divided into ***agricultural*** ones, those intended for use in the production sectors, in the storage and processing of agricultural products, on pastures and in planted forests. Another division of pesticides refers to ***non-agricultural*** pesticides, i.e. those intended for the protection of native forests, other ecosystems or water environments. Both must be registered by the Ministry of the Environment and IBAMA, in accordance with the guidelines and requirements of the Ministries of Agriculture, Livestock and Supply and Health. Examples of non-agricultural pesticides are those for use in urban and industrial, household, public or collective environments, for water treatment and for use in public health campaigns - whose registrations are granted by the Ministry of Health/ANVISA, in compliance with the guidelines and requirements of the Ministries of Agriculture and the Environment (MMA, 2010).

There are around fifteen thousand formulations for four hundred different pesticides, of which around eight thousand are licensed in Brazil, which is one of the five largest consumers of pesticides in the world (FEPAM, 2007).

According to Law 7.802, of July 11, 1989, there is a great diversity of products, approximately three hundred active ingredients in two thousand commercial formulations in Brazil, classified as: Insecticides, fungicides, herbicides, rodenticides, acaricides, nematicides, molluscicides and fumigants.

Insecticides act to combat insects, larvae and ants and belong to four chemical groups: Organophosphates, carbamates, organochlorines and pyrethroids.

Organophosphates: organic compounds derived from phosphoric acid, thiophosphoric acid or dithiophosphoric acid. E.g. Folidol, Azodrin, Malation, Diazinon, Nuvacron, Tamaron, Rhodiatox. Carbamates are derivatives of carbamic acid. Ex: Carbaril, Temik, Zectram, Furadan.

Organochlorines are carbon-based compounds with chlorine radicals, derived from cyclohexane or cyclodiene, which were widely used in agriculture as insecticides, but their use has since been restricted or banned. E.g. Aldrin, Endrin, DDT, Endosulfan, Heptachlor, Lindane, Mirex.

Pyrethroids are compounds that have structures similar to pyrethrin, some of the compounds are: alethrin, resmethrin, decamethrin, cypermethrin and phenpropanate. E.g. Decis, Protector, K-Otrine, SBP. Fungicides act to combat fungi and the main chemical groups are: Ethylene-bis-dithiocarbamates (Maneb, Mancozeb, Dithane, Zineb, Tiram); Triphenyl stannic acid (Duter and Brestan); Captan (Ortocide and Merpan) and Hexachlorobenzene.

Herbicides are used to combat weeds. In the last two decades, this group has seen increasing use in agriculture. Its main representatives are: Paraquat (Gramoxone); Glyphosate (Round-up); Pentachlorophenol; Phenoxyacetic acid derivatives (2,4 dichlorophenoxyacetic (2,4 D) and 2,4,5 trichlorophenoxyacetic (2,4,5 T)); The mixture of 2,4 D and 2,4,5 T represents the main component of Agent Orange, used as a defoliant in the Vietnam War (the commercial name for this mixture is Tordon); Dinitrophenols (Dinoseb) (LORENZI, 2006).

Pesticides are still classified according to their toxic power, which is fundamental for understanding the toxicity of a product. In Brazil, toxicological classification is the responsibility of the Ministry of Health. On pesticide labels, you can see that each one has a band indicating the toxicological class symbolized by colors, i.e. the degree of toxicity for a product that is extremely toxic is red. The higher the level of toxicity, the greater the dangers of poisoning to the worker and the environment, if the appropriate means of protection are not taken during handling (SOUZA, 2006; ZAMBRONE, 2002).

Pesticide products are divided into four categories in terms of their Toxicological Classification: Class I - Extremely Toxic Products; they have a red stripe; Class II - Highly Toxic Products; they have a yellow stripe; Class III - Medium Toxic Products; they have a blue stripe; Class IV - Slightly Toxic Products; they have a green stripe (BRASIL, 1998).

Agriculture is one of the human activities characterized as one of the main degraders of natural ecosystems, mainly due to the adoption of a production model based on monoculture and the intensive use of chemical inputs that cause biological imbalances and environmental contamination. Across the planet, 66% of arable land has already been degraded by monocultures, which destroy biodiversity and use large quantities of agrochemicals (CETESB, 2009).

Therefore, humans alter the patterns of watercourses and storage to build dams, control floods, drain wetlands and pump groundwater in order to achieve higher and more stable incomes (MCNEELY and SCHERR, 2009) through farming, mainly to meet the demand to supply the world food market.

1.1 - Impacts on the ecosystem

Due to the replacement of labour by mechanization in various agricultural activities and the use of pesticides from the 1930s onwards, as well as the use of biotechnology, genetically modified organisms (transgenics) were the main technological changes that contributed to altering the agricultural process (SILVA *et al.*, 2004; TRAPÉ, 1993).

In order to meet the need for good agricultural production, increasingly innovative technologies are being used, which often, even if they are beyond the farmer's control, cause environmental damage, such as threatening the survival of other species and even reducing the quality of life of the human species itself (Embrapa).

According to Thomson (2007) *and* Miller (1931), technologies significantly alter nature, depending on how they are used, for example:

1) The practice of deforestation can destroy, fragment, degrade and simplify wildlife habitats, reduce biodiversity, not to mention landfill wetlands to grow food or build buildings, highways, which represents a loss of overall biodiversity and degradation of the Earth's natural capital;

2) the misuse of an ever-increasing percentage of the land's productivity, which sustains all consumer species (including humans). This type of alteration is the main reason why the habitats and food stocks of an increasing number of species are being displaced or eliminated;

3) by strengthening some harmful species and disease-causing bacteria. The excessive use of pesticides and antibiotics has accelerated natural selection among rapidly reproducing populations of bacteria and pests and has led to genetic resistance to these chemicals;

4) by eliminating certain predators. Some farmers want to eliminate predators from their farms, thereby simplifying and damaging food chains and natural ecosystems;

5) This will happen if, deliberately or accidentally, new or non-native species are introduced into communities, where most of these species, such as grain for cultivation and domesticated livestock, are beneficial to humans, but some are harmful even to other species.

6) the use of some renewable resources faster than the natural replenishment capacity, for example ranchers allow cattle to graze excessively, until erosion turns these areas into less productive deserts;

7) by interfering with chemical cycling and normal energy flows in ecosystems. Soil nutrients can be depleted by monoculture, tree plantations, construction sites and other communities, such as lakes and coastal waters, thus promoting an increase in carbon dioxide emissions through the burning of fossil fuels, forests, pastures and deforestation. These and other increases from greenhouse effect activities can trigger global climate change by altering the flow of energy through the troposphere;

8) human-dominated systems are becoming increasingly dependent on the non-renewable energy of fossil fuels. Fossil fuel systems often generate pollution, release carbon dioxide into the atmosphere and waste energy;

To this end, the same author considers that the origin of the problem of pesticides is not only due to their misuse, but also to the practice of monoculture. The availability of unlimited quantities of certain foods encourages the multiplication of predators, commonly known as pests. The problem resulting from this process always tends to be the total destruction of the crop and a clash of populations between the "pests".

When pesticides are applied to crops, they are also transported to surface waters, and through various mechanisms, their residues can remain in the environment, causing damage to human health (IBANEZ *et al.*, 1996).

Pesticides accumulate in lipid tissues: liver, kidneys, brain and heart. Many foods are enriched in relation to the initial concentration of pesticides, which can be seen in the food chain (FARIA, 2004).

The degradation of natural capital for food production occurs as a result of the development of unbridled actions for excessively profitable agriculture. This implies a process of fauna imbalance, through the impact also caused by the use of agrochemicals, as well as deforestation, burning, leading to the pollution of soil, water, air, fauna and flora.

Entire populations have been seriously threatened by the impacts of agricultural projects, but amphibians are one of the most threatened, as they are extremely dependent on both aquatic and terrestrial environments.

The effects on the ecosystem resulting from the process of obtaining food are listed in Table 1.

Table 1: Main environmental effects in food production. Source: (TOWNSEND, 2006).

Loss of biodiversity	**Soil**	**Water**	**Atmospheric pollution**	**Human health**
Habitat loss and degradation due to the elimination of pastures and forests and the draining of wetlands	Erosion	Wasting water	Greenhouse gas emissions from the use of fossil fuels	Nitrates in drinking water
Death of fish due to pesticide run-off	Fertility fertility	Aquifer drainage	Other air pollutants from the use of fossil fuels	Pesticide residues in drinking water, food and air
Killing wild predators to protect domestic animals	Salinization	Increased run-off and flooding of land cleared for cultivation	Greenhouse gas emissions of nitrous oxide from the use of inorganic fertilizers	Contamination of drinking and bathing water by disease-causing organisms from pet waste
Loss of genetic diversity due to the replacement of wild crop lines by monocultures	Flooding	Sediment pollution due to erosion	Emission of methane, a greenhouse gas, by cattle	Contamination of meat by bacteria
	desertification	Death of fish due to pesticide run-off	Pollution from pesticide spraying	
		Surface water and groundwater pollution caused by pesticides and fertilizers		
		Over-fertilization of lakes due to fertilizer run-off, animal waste		

		from food processing.		

It can be pointed out that inadequate use and management of agricultural inputs degrades the soil, causes erosion and low productivity, thus contributing to the existence of floods, siltation, pollution of water sources and soil degradation, also causing losses to the producer. Inadequate soil management is an important degrading agent. The use of implements such as disc mechanisms compacts the soil, reduces biological activity in the medium and long term, destroying the surface structure and reducing plant production over time (MEDEIROS *et al.,* 2008).

According to a study carried out in the state of Tocantins, water contamination due to the use of agrochemicals occurs practically throughout the entire territory of Tocantins. This, among other major

environmental impacts resulting from anthropogenic actions diagnosed are shown in Table 2 (TOCANTINS, 2009):

Chart 2: Main Environmental Problems in the State of Tocantins. Source: NATURATINS, 2008.

Action	Spatial distribution	Main environmental impacts
Fires and Deforestation	North, east, central, south and central-west	Soil erosion; Increase in degraded areas; Contamination of watercourses by pesticides.
Disorganized mineral extraction	South East	Siltation of watercourses; Water and food chain contamination by mercury; Destabilization of the social base caused by disorderly migration to the mining areas; Economic impacts resulting from the waste of gold due to the lack of adequate technology and unequal competition between mines and companies, leading to a lack of significant investment in the mining sector.
Poaching and fishing	Midwest and Southwest	Reduction in the stock of some species of fauna.
Disorganized urban and industrial growth	Central	Contamination of watercourses, hindering public supply; Proliferation of vectors due to lack of basic sanitation; Increased incidence of water-borne diseases.
Suppression of legal reserve areas and riparian forests	North and Center	Erosion and siltation; Contamination of water bodies; scaring away wildlife.
Poor sanitation (waste and effluents)	The whole state	Hydric and edaphic contamination; Proliferation of water-borne diseases and those caused by insects and rodents.
Inadequate use of water resources	The whole state	Water shortages; Water contamination;

		Disease proliferation.
Indiscriminate use of pesticides	Northeast and Southwest	Hydric and edaphic contamination; Contamination of the food chain; Disease proliferation.
Inadequate land use	Southwest and Southeast	Erosion and siltation; Soil depletion; Loss of productivity.
Predatory tourism	Center and Midwest	Architectural damage; Water and soil contamination due to the accumulation of garbage; Indiscriminate hunting and fishing.

1.1.2 - Contamination of Water Resources

Pesticides are commonly applied to plants or to the soil, where around 50% of the total dose can reach the soil, regardless of how it is applied. From the soil it can be transported in three different ways: leaching, volatilization and surface runoff, thus contaminating the ecosystem (MURT and NAG, 1991; FARIA, 2004).

The dumping of toxic substances by agriculture is one of the sources of water contamination, as more and more insecticides, herbicides, fungicides and a whole series of pesticides are being used. Even if these substances reach rivers and lakes in small quantities, bioaccumulation will make their action highly damaging along food chains (FARIA, 2004).

The indiscriminate use of water resources is due to the increase in population and the lack of control over the impact of human activities on the natural environment, which is a cause for concern for experts as the total amount of water withdrawn from nature for human consumption has grown ninefold since 1950, with the freshwater reserve per person in the world falling from 16.8 thousand m^3 (in 1950) to 7.3 thousand (1998) and tending to fall to 4.8 thousand m^3 over the next 20 years. This is without taking into account the fact that consumption is uneven, an example of which is Brazil, where some regions are experiencing critical situations due to a shortage or lack of water (CAMARGO *et al.*, 2002).

For many years, the process of intensifying man's ability to intervene in the natural environment developed gradually and cumulatively, and for some time these changes were unnoticed, even saying that apparently they were not significant compared to today. That was until the first great scientific and technological revolution took place, which had an impact on the natural environment due to the increase in production capacity. For a long time, societies had been constantly on the move in search of bigger and better harvests, leading to the agricultural revolution, allowing people to settle and the first cities to emerge (DIAS, 2008).

According to the report "O Estado Real das Aguas no Brasil - 2003/2004" (The Real State of Water in Brazil - 2003/2004), drawn up by the Water Ombudsman's Office, water pollution has increased fivefold in the last ten years. The report was based on 35,000 complaints of damage to the environment and public civil actions that have already received a court ruling. It points out that the main source of contamination in the country comes from agro-industrial and industrial activities, which are responsible for the consumption of 90% of the water that is returned contaminated after use (DIAS, 2008).

Unfortunately, action is only taken in the form of a reaction (Polluter Pays Principle) and not preventively (Machado, 2003), i.e. action is only taken after some kind of damage has occurred. If this thinking is extended to contamination, it can be seen that some kind of decision was only taken after several environmental impacts had occurred. One of the acts that began to translate the effects of man's relationship with the environment was Rachel Carson's literary work "Silent Spring", in which she tried to alert people to react against the abuses of chemical pesticides, trying to awaken them to other ways of producing without harming them, starting to use biological control products, which consist of using fungi, bacteria and insects to combat the parasites that feed on plants.

Water pollution is the alteration of natural quality due to the degradation that takes place, resulting in damage or destruction of the ecosystem, making it unsuitable for human consumption, caused by pesticide residues and fertilizers from agriculture, where they are widely used, the excess of which is leached into bodies of water or percolates reaching underground reservoirs (IBGE, 2002).

According to a survey carried out by IBGE (2005), the second biggest cause of water contamination in Brazil is due to the use of pesticides and fertilizers, second only to the dumping of domestic sewage. For 2009, the soybean plantation area increased by 2.0% compared to 2008. Consequently, this increase will also lead to an increase in the amount of pesticides being used.

The IBGE (2002) carried out a survey on the most frequent environmental impacts in Brazilian municipalities. Figure 1 shows the existing impacts in Brazil, where pollution of water resources was in second place.

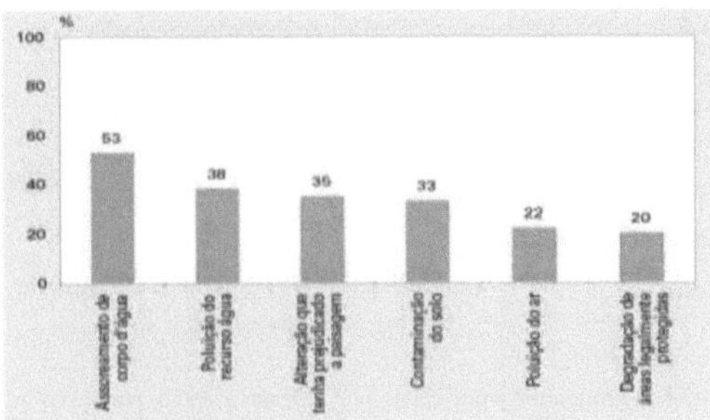

Figure 1: Environmental impacts related to Brazil, 2002.

Source: IBGE, 2002.

Contamination by agrochemicals (pesticides and fertilizers) is a topic of study that is attracting increasing attention, given its consequences for human health and the risk of environmental degradation caused by these products. Water pollution, for example, leads to a whole series of impacts related to the impossibility of practicing certain entertainments (bathing and sporting activities), the use of (drinking) water and the loss of species dependent on this habitat, such as fish (IBGE, 2002).

1.2 - Legal framework on pesticides

Brazil plays an important role in the global production of pesticides. As a result, the demands for compliance with legal standards are increasing, even internationally, due to the risk of toxicological and environmental contamination. The most important legal reference is Federal Law No. 7802/89, which governs the process of registering a pesticide product, regulated by Decree No. 4074/02. However, many people are still unaware of the existing environmental legislation in this regard.

In order to be produced, exported, imported, marketed and used, pesticides must first be registered with a federal agency, in accordance with the guidelines and requirements of the federal agencies responsible for the health, environment and agriculture sectors. IBAMA registers the non-agricultural pesticides mentioned above and assesses the potential for environmental danger of all pesticides registered in Brazil (MMA, 2010).

According to Law 7.802/89, article 3, paragraph 6, the registration of pesticides is prohibited in Brazil: for which Brazil does not have methods for deactivating their components, in such a way as to prevent their remaining residues from causing risks to the environment and public health; for which there is no effective antidote or treatment in Brazil; which reveal teratogenic, carcinogenic or mutagenic characteristics, according to the up-to-date results of experiments by the scientific community; which cause hormonal disorders, damage to the reproductive system, according to up-to-date procedures and experiments in the scientific community; which prove to be more dangerous to humans than laboratory tests on animals have been able to demonstrate, according to up-to-date technical and scientific criteria; and whose characteristics cause damage to the environment.

Table 3 lists the public policies aimed at controlling the use of pesticides in Brazil that contribute directly or indirectly to the environmental conservation of the existing biome.

Chart 3: Laws controlling the use of pesticides in Brazil, 2009. **Source:** MMA and SRHMA/TO *website*, 2009.

FEDERAL LAW	STATE LAW
- *Law No. 7.802, of July 11, 1989,* provides for research, experimentation, production, packaging and labeling, transport, storage, marketing, commercial advertising, use, import, export, final destination of residues and packaging, registration, classification, control, inspection and inspection of pesticides, their components and the like, and makes other provisions. - Decree No. 4.074 of 04.01.2002, regulates Law No. 7.802 of July 11, 1989, which provides for research, experimentation, production, packaging and labeling, the	- Law No. 261, of February 20, 1991, provides for the Environmental Policy of the State of Tocantins; - Decree No. 10.459, of 08/06/94 Regulating Law No. 261, of 20/02/91; - Law No. 771, of July 7, 1995, provides for the Forestry Policy of the State of Tocantins;

transportation, storage, marketing, commercial advertising, use, import, export, final destination of residues and packaging, registration, classification, control, inspection and inspection of pesticides, their components and the like, and makes other provisions. - *Law No. 9.974, of June 6, 2000,* amends Law No. 7.802, of July 11, 1989, which provides for research, experimentation, production, packaging and labeling, transport, storage, marketing, commercial advertising, use, import, export, final destination of residues and packaging, registration, classification, control, inspection and inspection of pesticides, their components and purposes, and makes other provisions. - *Law No. 9.605, of February 12, 1998*, provides for criminal and administrative sanctions arising from conduct and activities harmful to the environment, and makes other provisions. - *CONAMA Resolution 020, of June 18, 1986*, which deals with the classification of water uses and establishes maximum levels for potentially harmful substances. - IBAMA Normative Ordinance No. 84 of October 15, 1996.	- Decree No. 838 of 13/10/1999 Regulates Law 771 of 07/07/95; - Law No. 224, of December 26, 1990, on pesticides and other provisions. - Law No. 1.307, of March 22, 2002, on the State Water Resources Policy.

1.3 - Biological aspects of *Leptodactylus* as a bioindicator

Amphibians have a complex life cycle and a greater variety of reproductive modes than any other vertebrate (POUGH *et al.*, 1999). They are very sensitive and vulnerable to environmental variations, mainly due to the permeability of their skin and the fact that their life cycle depends on both the aquatic and terrestrial environments. It is therefore considered to be a bioindicator of environmental quality (FEIO *et al.*, 1998).

Modern *amphibians* form a monophyletic group well corroborated by some synapomorphies: pedicellate teeth, the presence of a sensory region in the inner ear known as the papilla *amphibiorum*, the presence of a bone in the ear known as the operculum, the presence of venom glands in the integument and fat bodies situated cranially to the gonads. They are included in three orders that make up the subclass Lissamphibia: Gymnophiona, Urodela and Anura. The reproductive diversity of the group is enormous and although there are exceptions, most species have a larval life stage - terrestrial or aquatic - which, after metamorphosis, gives rise to the adult animal.

Anurans (frogs, toads and tree frogs) are adapted for jumping and do not have tails as adults. Their caudal vertebrae are fused in the adult in a bone known as the urostyle. There are approximately 4,000 species, divided into around 20 families, whose classification is controversial. In most anurans, fertilization is external, but there are ovoviviparous and viviparous species with internal fertilization. Of all the vertebrates, anurans have the greatest diversity of reproductive modes, being able to use practically every type of aquatic environment available and in some cases do without them (HOFLING, 1995).

The species *Leptodactylus latrans* was first described by *Linnaeus* in 1958. It became known as the butter frog, giote or caçote and can reach up to 15 centimeters in rostro-cloacal length and weigh up to 350 grams. It is widely distributed throughout South America and is very common in streams and lagoons throughout the Northeast. They feed on small arthropods and vertebrates such as other frogs, fish and snakes. They have a vocalization reminiscent of a dog's bark, but hoarser and weaker (FREITAS, 2007).

1.3.1 - Anatomical and histological aspects

Rice (Oryza sativa L.) is one of the oldest cultivated species in the world, occupying approximately 10% of arable land. With the increasing reduction in the areas available for production, the search for new alternatives that reconcile intensive cultivation with high yields has become indispensable (EMBRAPA, 2004). The most damaging effects of pesticides occur after chronic exposure to irrigated rice (CONSTANTINO, 2007).

Various biochemical parameters have been tested in fish in relation to their responses to toxic substances. Among the most widely investigated indicators in these animals are the enzymes present in liver tissue, involved in the detoxification of xenobiotics and their metabolites (LECH & VODICNIK, 1985), and their use is amplified by their response to toxicants in a similar way to that of large vertebrates (AL-SABTI; METCALFE, 1995), and also because of their great affinity with the toxic agent due to their high bioaccumulative power, which also makes them more susceptible to possible poisoning, generating multiple serious alterations with consequences for populations, communities or ecosystems, depending on the degree and time of exposure (ARIAS, 2002).

The behavior of amphibians in the face of an altered environment generates information that makes it possible to assess the environmental stress to which the population is subjected, which makes it an excellent bioindicator of environmental quality, as many studies have shown (ABILHOA *et al.*, 2003).

Using histology on animal tissues, it is possible to assess changes suffered as a result of aggression or stress caused by the environment in which they live.

Using a perpendicular cut in the skin of a frog, you can see its different layers. In the epidermis, the germinal and horny layers can be seen. In the dermis, at higher magnification, the spongy and compact strata. Deep in the more superficial spongy strata, there are many glands of epidermal origin. The mucous glands have amorphous contents and the serous glands are granular. The compact stratum of the dermis is made up of bundles of collagen fibers, parallel to the surface of the integument (HOFLING, 1995).

In this class of vertebrates, the integument shows adaptive modifications to terrestrial life. For example, the superficial layers of cells are more or less keratinized, and in amphibians more adapted to terrestrial life, such as frogs, the cells in this layer represent a considerable deposit of keratin. Amphibians have developed pluricellular glands in the integument (GEORGE and CASTRO, 1998).

The dermis of amphibians is made up of two distinct layers. The outer layer is made up of loose connective tissue (papillary dermis), containing vessels, cutaneous nerves and chromatophores. There is little adipose tissue. The chromatophores are of three types, depending on the nature of the pigment they contain: melanophores, with melanin; xanthophores containing yellowish pigment, lipochrome; guanophores, containing guanine crystals. The deep layer of the dermis, although thinner than the outer layer, is made up of one or more layers of loose connective tissue, which is in contact with the underlying skeletal muscles (GEORGE and CASTRO, 1998).

The use of agrochemicals is a practice in Brazilian agriculture which has strongly contributed to the degradation of ecosystems. The soil is where organochlorines are most concentrated, as they have a longer half-life and are transferred to future crops, or can remain active for a short period of time, preventing the establishment of new plant populations. Water is most often affected by the dumping of leftover formulations, washing equipment and, above all, direct application and rainfall (RUEGG *et al.*, 1991).

Lipid peroxidation, assessed through the formation of thiobarbituric acid reactive species (TBARS), was not altered during the first stages of rice production, between November and December 2006. However, during the rice harvest, in April, there was a significant increase in the concentration of TBARS, demonstrating greater damage from the pesticide after chronic exposure to fish, due to its bioaccumulative effect (CONSTANTINO, 2007).

After analyzing the data obtained from different collections of fish liver tissue, used to assess oxidative damage following the use of pesticides in irrigated rice cultivation, there was a significant increase during harvest in both lipid peroxidation and protein carbonylation (CONSTANTINO, 2007).

Even at concentrations below the CL50, the herbicide is capable of causing damage to the liver of exposed fish (ALBINATI *et al.*, 2007).

There are hundreds of pollutants that affect the aquatic environment, the effects of which are worrying. Aquatic pollution is commonly associated with the discharge of domestic, industrial or agricultural effluents. In agricultural areas, the leaching of surface water and the infiltration of interstitial water into rivers and lakes can introduce nutrients (from fertilizers) and pesticides in substantial quantities into these bodies of water (MARTINEZ; CÓLUS, 2002).

Often, even in non-lethal aquatic concentrations, pesticides affect the structure and function of natural communities. In the aquatic environment, pesticides cause impacts at multiple levels, including molecules, tissues, organs, individuals, populations and communities (GRISOLIA, 2005). Fish are relatively sensitive to changes in the environment and the toxic effects of pollutants can be evident at the cellular and tissue level, before significant changes in behavior or external appearance can be identified (VAN DYK, 2005).

Histopathological analysis of the fish liver showed moderate vacuolization in the group exposed to 6.67 ppm of Roundup® and slight vacuolization in the group exposed to 13.3 ppm, as well as intense congestion of vessels and sinusoids in the groups exposed to the herbicide.

Congestion of sinusoids, in addition to the onset of fibrosis, was an alteration also reported by NESKOVIC *et al.* (1996), in carp (*Cyprinus carpio*) exposed to 10 mg/L of glyphosate for 14 days. LANGIANO (2006), in addition to other alterations, also found liver congestion in curimbatas exposed to 7.5 and 10.0 ppm of Roundup® for 6, 24 and 96 hours.

Although hepatocytes are generally full of glycogen or neutral fat, if food is sufficient (ROBERTS, 1981), three situations can occur during the stress response phases: a) a mobilization of hepatic glycogen in the alarm phase; b) in the resistance and adaptation phase, glycogen can be low or not; c) in the exhaustion phase there is a depletion of hepatic glycogen (HEATH, 1995). The discrete vacuolization found in the liver of the animals in the group exposed to 13.3 ppm of Roundup® suggests a mobilization of glycogen due to chemical stress, which is considered low toxic, according to IBAMA classification standards.

Despite this, fish exposed to the herbicide at concentrations of 6.67 and 13.3 ppm showed hepatic alterations consistent with an imbalance in their homeostasis.

Experiments carried out with six increasing concentrations (0.0; 0.5; 1.0; 1.5; 2.0 and 2.5 mL) of formaldehyde, one of the most commonly used chemicals in aquaculture, showed that after 96 hours of exposure, the surviving larvae from each treatment were collected. The larvae were fixed in *bouin* solution and embedded in plastic paraffin to obtain 5 µm histological sections and stained with hematoxylin/eosin (CRUZ *et al.*, 2005).

Aqueous formaldehyde solution is widely used as an antifungal and anti-parasitic agent (RACH *et al.* 1997), mainly for protozoa and monogeneans. Its use has been approved in the United States by the *Food and Drugs Administration*

(FDA) for use in the control of fungi, protozoa and external parasites (FDA, 1992 and 1998).

According to MARTINS (2004), 37% formaldehyde can be used in short-term baths (up to 60 minutes) at a concentration of 150 to 250 mL/m^3 and in long-term baths (24 h) at a concentration of 10 to 15 mL/m^3 . However, he points out that formaldehyde damages the gill epithelium and reduces the concentration of dissolved oxygen in the water. NOGA (1996) also observed that formaldehyde damages the gills and that its toxicity increases in acidic and high-temperature waters.

Formaldehyde was considered to be moderately toxic to trawler *larvae* (*H. lacerdae*), based on the classification by ZUCKER (1985), with LC 50 for fish and aquatic invertebrates ranging from 1 to 10 mg/L.

The liver of the trawler larvae in the control and 0.5 mL/L treatments showed a cord-like organization of the hepatocytes, forming a plate of cells in direct contact with the sinusoids. The hepatocytes were hexagonal in shape, with a central, basophilic nucleus, decondensed chromatin and a visible nucleolus. The pink cytoplasm indicated high acidophilia and the presence of some glycogen granules. The sinusoids were lined by a thin endothelium with small endothelial cells and prominent nuclei, as well as the presence of blood cells inside the sinusoids (CRUZ *et al.*, 2005).

According to ROBERTS (1981), fish differ from mammals in not always having hepatocytes in a cordonal arrangement or in hepatocyte lobes. However, in this study it was possible to see the cordonal arrangement in trawler larvae. Similar histological characteristics were also described by SOUZA *et al.* (2001) in pacu (*P. mesopotamicus*).

In the treatments with 1.0, 1.5 and 2.0 mL/L, there was cordonal disruption of the hepatocytes, their nuclei showed condensed chromatin and a reduction in size, as well as cell fusion in some regions and congestion in the sinusoids after 96 hours of exposure. The changes that occurred in the trawler's liver were similar to those described by DUTTA *et al.* (1993) for *Heteropneustes fosilis* exposed to malathion for 96 hours, by RODRIGUES and FANTA (1998) for *Brachydario rerio exposed to* dimethoate for eight hours and by FANTA *et al.* (2003) for *Corydoras paleatus* exposed to methyl parathion.

According to these authors, organophosphate insecticides caused changes in the cytoplasmic granulation of hepatocytes, reducing the efficiency of the liver and leading to malfunctioning of other organs in the fish, which can cause death. These alterations may indicate the initial degeneration of the hepatocytes' intracellular processes, as the liver performs important functions in the animal's detoxification process (DUTTA *et al.* 1993; HASCHEK and ROUSSEAUX, 1996).

According to JHONSON *et al.* (1993), changes in the histology of the organs appear over a medium period of time as a response to sublethal stress. Therefore, as the detoxification mechanism of formaldehyde in fish is not known, it is necessary to test other exposure times and concentrations for the trawler to assess the occurrence of irreversible alterations in the liver (MONTENEGRO and FRANCO, 1999).

Exposure to sublethal concentrations of formaldehyde at 10%/L of water caused minor alterations in the liver of trawler larvae, but care must be taken when using formaldehyde to treat parasites due to its acute toxicity (CRUZ *et al.*, 2005).

1.4 - Bioindicators

Bioindicators are species or higher taxonomic groups with characteristics (such as presence/absence; population density, dispersal, reproductive success) that can theoretically be used as an index for other ecosystem attributes that are more difficult or expensive to measure (LANDRES, 1992).

Several amphibian species have a wide distribution and could potentially serve as key species for assessing long term geographical or global changes in the environment. Other species are habitat specialists or have a restricted distribution, and can indicate a local disturbance (HEYER *et al.*, 1993). The ability of this group to adapt to the various structural manifestations of habitats results in various endemisms. Extinctions and population declines have occurred for reasons that are not very well understood, but appear to be influenced by deforestation in several cases (HADDAD and ABE, 1999).

Designating a species as an indicator requires prior knowledge of its community structure, including the species' life cycle, seasonality and natural variations, so that it is possible to compare conditions before and after an environmental disturbance (RAZ, 2000).

According to LÉVÊQUE (1996) there are two main categories of bioindicators: bioaccumulation indicators which fix pollutants and effect indicators which can be used at different levels: biochemical, physiological or ecological. Ecological indicators can be negative (they regress with pollution) or positive (they develop in a polluted environment).

The two most important characteristics of bioindicators are: a) they make it possible to identify the interactions that occur between contaminants and living organisms; b) they make it possible to measure sub-lethal effects, which makes it possible to implement preventive actions, hence the importance and current interest in incorporating bioindicator analysis into environmental contamination assessment programs (ARIAS *et al.*, 2002).

The main requirements that an organism must present to be a bioindicator are: established taxonomy; known biology and natural history; it must be easily observable and manipulated in the field and in the laboratory; wide worldwide distribution (ecologically well diversified); present specialized species in restricted habitats (sedentary within an ecosystem); pattern of species richness widely correlated with other taxa; high ecological sensitivity and fidelity; they must be abundant and; possess species with economic potential (CODDINGTON *et al.*, 1996).

In this sense, amphibians are also considered ideal models for studies on the effects of ecosystem fragmentation, mainly due to their low mobility, physiological requirements, habitat specificity and ease of study (SILVANO *et al.*, 2003). They have a complex life cycle and exhibit the greatest variety of reproductive modes and life history of any vertebrate group (DUELLMAN and TRUEB, 1986; POUGH *et al.*, 1999).

Various environmental factors can affect the size of populations of organisms, but specific characteristics of amphibians, such as skin permeability and a life cycle dependent on both aquatic and terrestrial environments, make these terrestrial vertebrates more vulnerable to environmental variations. Consequently, the reduction in the amphibian population in a given region is considered a bioindicator of environmental quality (RABB, 1990; STEBBINS & COHEN, 1995; FEIO *et al.*, 1998).

Recently, several studies have detected a worldwide decline in amphibian populations (RABB, 1990; VIAL, 1991). One of the main factors that may be causing this decrease and affecting the development of eggs is deforestation, which causes a reduction in vegetation cover, eliminating the specific microenvironments of amphibians, reducing the rate of individuals in the populations (HADDAD, 1998; POUGH *et al.*, 1999).

The authors HADDAD and BASTOS (1997) suggest that male amphibians are more exposed to predators during the active search for females. Studies of the amphibian fauna in the Cerrado are still scarce (TOCANTINS, 2003). The anuran species of the Cerrado have at least one of the following characteristics: reproduction occurs in the rainy season, individuals form dense aggregations in bodies of water, they are long-lived and territorial. In addition, they are vulnerable to contamination, as their skin is highly permeable, and they depend on both the aquatic and terrestrial environments (DIXO, 2001).

Amphibians are considered excellent bioindicators, not least because of their close relationship with microenvironments, wet environments and high sensitivity to environmental disturbances, and are considered important in monitoring studies in conservation units (TOCANTINS, 2003). According to CARDOSO *et al.* (1989), leptodactylids suffer less water loss due to contact with wet soil.

1.5 - Amphibians in Brazil

Brazil is the country with the greatest biodiversity of amphibians on the planet, with more than 600 species of anurans. Of these species, 60% are endemic, which emphasizes the need for their preservation. These creatures are the group facing the most serious problems caused by the intensive and unsustainable use of natural resources. Several anuran species are suffering population declines, mainly due to the growing search for areas for agricultural expansion (TOMMAZI, 2008).

Although the Brazilian fauna is very large, the amphibian population worldwide is disappearing at an alarming rate, which may be caused by regional or global effects of human activities that can probably affect other organisms (AURICCHIO and SALOMÂO, 2002).

In Brazil, the main threat to amphibian conservation is the destruction of their habitat as a result of deforestation, the advance of the agricultural frontier, mining, fires and the development of infrastructure and urbanization. Little is known about other causes of amphibian decline observed worldwide, such as the effects of pesticides (SILVANO and SEGALLA, 2005).

1.6 - Family Leptodactylidae

The name *Leptodactylus latrans* has been used for various populations distributed throughout South America, east of the Andes. The name comes from the spots on its back, which resemble ocelli. It has at least six dorsolateral folds and is sexually dimorphic. The males are larger and have two spines (black or white) of epidermal origin at the site of the pre-plexus, they have robust arms and are appreciated by many people as food. The females have a callus instead of spines. They are large to very large frogs (between 90 and 110 mm CRC) that prey on other amphibian species. They spawn in foam nests on the surface of the water. The females take parental care of both the eggs and the tadpoles. They often open channels, connecting pools, thus preventing the tadpoles from becoming trapped and dying from desiccation (HADDAD, 2005).

Figure 2 shows the geographical distribution of the *Leptodactylidae* family throughout the Americas, according to the highlighted area.

Figure 2: Geographical distribution of the *Leptodactylidae* family in the Americas. Source: (MACHADO, 2009).

1.7 - Amphibians in Tocantins

There are still few studies on amphibians in the state of Tocantins. However, the region has a very rich biodiversity, taking into account Lagoa da Confusao and Cristalândia - TO.

According to CHEISA (2007), the number of species (richness) of amphibians found in "ipucas" was higher than in agricultural areas, which is probably due to the greater environmental heterogeneity of these areas. Animals were identified in the region, all of which were distributed in 4 families: Bufonidae, Hylidae, Microhylidae and Leptodactylidae.

Due to the number of agricultural projects that are developed every year, which use rivers for irrigation and large quantities of various pesticides, the balance of fauna and flora is put at risk.

1.8 - Historical Aspects

Tropical regions: municipalities of Lagoa da Confusao and Cristalândia - TO

According to TOCANTINS (2008) Lagoa da Confusao is located in the western mesoregion of the state of Tocantins, on the right bank of the Araguaia River, on the border with Mato Grosso. The seat of the municipality is located at geographic coordinates 10°47'37" S latitude and 49°37'25" longitude. Its average altitude is 184 meters above sea level. It is 195 km from Palmas, passing through the town of Paraiso do Tocantins.

Lagoa da Confusao, created in 1993, has an area of 10,565 km^2 and according to IBGE data (2007) has 8,220 inhabitants. Finally, the third municipality studied was Cristalândia, created in 1953. It has an area of 1,848 km^2 , a population of 6,520 inhabitants and is 53 km from the municipality of Lagoa da Confusao. The municipality of Lagoa da Confusao borders the following municipalities: Pium to the north, Cristalândia, Dueré and Santa Rita do Tocantins to the east, Formoso do Araguaia to the south and Mato Grosso to the west. It is part of the Araguaia Tourism Corridor, created by the Tocantins State Government to take advantage of the region's potential for sustainable tourism development. The municipality is part of the Araguaia National Park and part of the Araguaia Park Indigenous Land, both on Bancloacal Island (TOCANTINS, 2008).

The region's climate is characterized by two well-defined seasons: a dry one (winter) and a rainy one (summer). The average annual temperature is around 24°C (BRASIL, 1994). The highest temperatures occur in September, with temperatures of around 35°C, and the lowest in July, with an average temperature of 15°C. The maximum temperature of the year is 42°C and the absolute minimum is 8°C. The annual rainfall total is around 1,750 to 2,400 mm, with the wettest months being from November to March (TOCANTINS, 1997). The relative humidity is considered high, at around 87.84%. In the drier months, humidity ranges from 79.6% in August to 88.65% in April. In the wetter months, humidity ranges from 93.99% in January to 90.64% in October (BRASIL, 1994; TOCANTINS, 1997).

The relief is classified as Accumulation Form, i.e. River Terraces, River Plateaus and Floodable Accumulation Areas. They are reliefs resulting from the deposit of sediments in fluvial, paludal and lacustrine regions, normally subject to flooding (TOCANTINS, 2008).

The slope is equal to or less than 5%: there is a predominance of areas with gentle slopes where, in most soils, surface runoff is slow or medium. The slope itself does not prevent or hinder the work of any type of agricultural machinery. Water erosion doesn't pose any major problems. On some types of soil, however, simpler conservation practices are recommended. For highly erodible soils with very long slope lengths, complex practices may be necessary, such as terrace systems and retention strips (SEPLAN, 2008).

The region's soils are predominantly concretionary with 22.8% or 63,468.1 km^2 . As for their erodibility, it can be said that it is considered to be slightly weak, which includes areas formed by soils ranging from well to strongly drained. They are deep soils and occur on gently undulating terrain (predominantly slopes between 3 and 8%). The ecodynamics of the landscape vary from stable to transitional (pedogenesis > morphogenesis). Surface runoff processes are diffuse and slow, with occasional concentrated flows - 110,477.8 km^2 (39.7%) (TOCANTINS, 2008).

The phytoecological regions of the region are basically covered by Cerrado, where open xeromorphic vegetation predominates, dominated and marked by a herbaceous layer. It occurs almost everywhere in the state, preferably in seasonal climates (more or less 6 dry months), and is also found in ombrophilous climates, when it must cover leached and/or aluminized soils (TOCANTINS, 2008).

With regard to potential land use, the region is characterized by the predominance of areas for semi-intensive livestock farming and/or forestry, which corresponds to 5.1% or 14,291.3 km^2 , as well as agricultural land - rainfed agriculture, irrigated by central pivot and flooding, forestry areas (2,784.2 km^2 or 1%). The vegetation present is cerrado of the campo, campo cerrado, cerradâo and campo parque types (157,373.4 - 56.6%) (TOCANTINS, 2008).

With regard to restricted legal use and the potential for environmental conservation, it is part of an area where sustainable use must be respected (TOCANTINS, 2008), which according to TOCANTINS (2005) means that these are areas where the environment can be exploited in such a way as to guarantee the continuity of renewable environmental resources and ecological processes, maintaining biodiversity and other ecological attributes. There are tourist attractions such as beaches, sport fishing and caves (TOCANTINS, 2008).

1.9 - Impact of pesticides on amphibians

It even affects local fauna by altering the composition of crops, which are the basis of their diet, affecting everything from predators to pollinators and dispersers, generating an ecological imbalance at all trophic levels. It therefore affects the dynamics of the food chain until it reaches humans, which can cause biochemical damage during chronic exposure (RUEGG *et al.*, 1991).

Pollution of aquatic ecosystems can lead to a loss of biodiversity, resulting in the reduction or disappearance of various populations. Pollution at the population level has degraded fundamental ecosystems through molecular alterations in fish, resulting in a reduction in the quality and sustainability of these ecosystems. Contamination of aquatic resources at the organism level is a source of human concern, given that direct and indirect consumption of contaminated fish and water can cause serious damage to the organism (RAMSDAF, 2007).

The disappearance of several species of amphibian may indicate a decline in environmental quality in various parts of the world. Amphibians (frogs, toads and salamanders) live part of their lives in water, part on land and some are classified as indicator species. Frogs, for example, are particularly vulnerable to environmental disturbances at various stages of their life cycle. Frogs as tadpoles live in water and eat plants, while as adults they live most of the time on land and eat insects, which can expose them to pesticides. Frog eggs have no protective shell to block ultraviolet (UV) radiation or pollution. As adults, they absorb air and water through their thin, permeable skins, which can immediately absorb pollutants from water, air and soil (THONSON, 2007 *apud* MILLER, 1931).

The same author lists a number of factors that can affect frogs and other amphibians at various times in their life cycles, including Habitat loss and fragmentation, especially from drainage and landfill of inland wetlands, deforestation and development; Prolonged drought, which dries up breeding pools, allowing few tadpoles to survive; Pollution, especially from exposure to pesticides, which can make frogs more vulnerable to bacterial, viral and fungal diseases and cause sexual abnormalities; Increase in ultraviolet radiation caused by the reduction of ozone in the atmosphere (which can harm young amphibian embryos in shallow ponds); parasites; over-hunting; viral and fungal diseases; Natural immigration or deliberate introduction of non-native predators and competitors (such as fish) and pathogenic organisms.

The extinction of amphibians is worrying for three reasons: 1) This trend suggests that environmental health is deteriorating in certain parts of the world, as amphibians are biological indicators that are sensitive to changes in environmental conditions, such as habitat loss and degradation, pollution, exposure to ultraviolet radiation and climate change. 2) Adult amphibians play important ecological roles in biological communities. For example, they eat more insects than birds, including mosquitoes. In some habitats, the extinction of certain amphibians could lead to the extinction of other species, such as: reptiles, birds, aquatic insects, fish, mammals and other amphibians that feed on them or their larvae. 3) Amphibians represent a genetic storehouse of pharmaceutical products waiting to be discovered. Compounds from amphibian skin secretions have been isolated and used as cloacalgesics and antibiotics, and as a treatment for burns and heart disease. The state of some amphibian indicator species is a warning sign (THONSON, 2007 *apud* MILLER, 1931).

The fact is that insects cannot be eliminated, as farmers would like to think. Eliminating one species has consequences for others, with the former becoming more resistant. What we can hope for is to maintain the ecological balance between them in order to eliminate these explosions. A constant low level of pests is preferable to alternating slightly better harvests with serious disasters. DDT is carcinogenic, affects the nervous system, the liver and also has an effect on fertility. Chemical substitutes for DDT are dangerous, but until a superior product is produced, they are still considered the best (TAYLOR, 1978).

The fact that their skin is permeable and their life cycle is biphasic makes amphibians dependent on water, which makes them sensitive to changes in the physico-chemical parameters of the water and the structure of the vegetation in the vicinity of water bodies. This can be harmful when pesticides are used on crops near the headwaters or when small dams and weirs are built for agriculture and livestock. In this way, man-made interventions alter the habitat of various animal species, causing species to disappear. Changes caused by environmental fragmentation can affect amphibian populations due to exposure to wind and sun and can lead to a decrease in humidity. The diversity of environments in a fragment can also favor the existence of a high number of amphibian species (SILVANO *et al.*, 2003).

In fragments where cattle are present, changes occur in the grass layer due to the replacement of native species by exotic ones, which cause shading of the soil and the disappearance of various species of reptiles and amphibians (SILVANO *et al.*, 2003).

This study aims to evaluate the contamination of *Leptodactylus latrans,* caused by the presence of pesticides, in the Urubu River sub-basin - TO. To this end, it has specific objectives which have been transformed into chapters as follows:

- Chapter II describes the quantitative and qualitative survey of pesticide producers and commercial establishments that were approached;

- Chapter III deals with the histological analysis of the integument, liver and reproductive system of *Leptodactylus latrans,* caught at three study sites, Cristalândia and Lagoa da Confusao - Tocantins.

❖ BIBLIOGRAPHICAL REFERENCES

ABILHOA, V. and DUBOC, L. F. Fish. In: MIKICH, S. B and BÉRNILS, R.S. (ed). **Red Book of Endangered Animals of Paranâ.** Curitiba: Mater Natura and Instituto Ambiental do Paranâ. 2003.

ALBINATI, A.C.L.; MOREIRA, E.L.T.; ALBINATI, R.C.B. **Acute toxicity of Roundup® herbicide to piauçu (*Leporinus macrocephalus*).** Rev. Bras. Saùde Prod. Anim., v.8, 2007, p.184-192.

AL-SABTI, K.; METCALFE, C.D. Fish micronuclei for assessing genotoxicity. In: water. **Mutation Research**, v.343, 1995. p. 121-135.

ARIAS, A.R.L.; VIANA, T.A.P.; INÂCIO, A.F. **Utilização de Bioindicadores como ferramentas de monitoramento e avaliação ambiental**: o caso de recursos hydricos. 2002. Available at: < http://www.ebape.fgv.br/radma/doc/FET/FET-020.pdf>. Accessed on: July 8, 2008.

AURICCIHIO P.; SALOMÂO M. da G. **Procedure for collecting samples of *L. latrans:*** Vertebrate collection and preparation techniques. 2002.

BRAZIL. National Electric Energy Agency. Available at http://www.aneel.gov.br. Accessed on October 3, 2008.

BRAZIL. **Ministry of Agriculture. Legislaçâo federal de agrotóxicos e afins** - Brasilia:, Departamento de Defesa e Inspeçâo Vegetal, 184p. 1998.

BRAZIL. Ministry of the Environment and Legal Amazon. Brazilian Institute for the Environment and Renewable Natural Resources - IBAMA. **Emergency Action Plan for the Araguaia National Park**. Brasilia: 1994. p. 56 (Basic information document).

BRAZIL. Ministry of the Environment. **First National Report to the Convention on Biological Diversity** - Brazil. Brasilia: MMA, 1998. 283p.

CAMARGO, A.; CAPOBIANCO, J. P. R.; OLIVEIRA, J. A. P. de.; Meio ambiente Brasil: avanços e obstàculos pós-Rio-92. Sao Paulo: Estaçâo da Liberdade: Instituto Socioambiental; Rio de Janeiro: Fundaçâo Getùlio Vargas, 2002.

CARDOSO, A.J., ANDRADE, G.V. and HADDAD, C.F.B. Spatial distribution in amphibian (Anura) communities in southeastern Brazil. Rev. Bras. Biol. 49(1):241-249. 1989.

CERRADO RURAL. Tocantins: seed quality and health are proven by Embrapa. Cerrado Rural: Agribusiness, the magazine of Cerrado Rural. October, 2009.

CETESB. **Environmental remediation**. PROSEAD Distance learning. Brasilia, 2009.

CHIESA, V. B. **Identification of the environmental quality of natural forest fragments - ipucas in hydro-agricultural and remnant areas, in the municipality of Lagoa da Confusao - TO, by means of bioindicators: anurans**. Palmas: UFT, 2007. 89 p.

CODDINGTON, J. A., YOUNG, L. H.; COYLE, F.A. Estimating spider species richness in a southern Appalachian cove hardwood forest. **J. Arachnol**. v. 24, p. 111-128. 1996.

CONSTANTINO, R.; *et al.* Natural Causes. In: RAMBALDI, D. M; OLIVEIRA, D. A. S. (eds.). **Fragmentation of Ecosystems: Causes, effects on biodiversity and public policy recommendations**. Brasilia: MMA/SBF, 2007. p. 24 - 40.

CRUZ, C. **Toxicological aspects of methyl parathion and aqueous extract of dried neem leaves (Azadirachta**

indica) for pacu (*Piaractus mesopotamicus*) and efficacy in the control of *Monogenea Dactylogyridae*. 2005. Thesis (Doctorate in Aquaculture)-Universidade Estadual Paulista, Jaboticabal, 2005.

D'AMATO, C., TORRES, J. P. M. MALM, O. DDT (Dichlorophenyltrichloroethane) Toxicity and environmental contamination - a review. **Quimica Nova,** Sao Paulo. v. 25, n. 6, p.995 - 1002. 2002.

Dixo, M. B. O. 2001. **Effect of forest fragmentation on the community of frogs and burrowing lizards in southern Bahia.** Sao Paulo. Master's dissertation - University of Sao Paulo.

DIAS, R. **Gestao Ambiental**: responsabilidade social e sustentabilidade. 1. ed. - 3. reprint - Sao Paulo: Atlas, 2008.

DUBOIS, J. C. L. **Agroforestry systems**. Executive Secretary of REBRAF. Article published in CAMARGO, A.; CAPOBIANCO, J. P. R.; OLIVEIRA, J. A. P. de.; Meio ambiente Brasil: avanços e obstàculos pós-Rio-92. Sao Paulo: Estaçâo da Liberdade: Instituto Socioambiental; Rio de Janeiro: Fundaçâo Getùlio Vargas, 2002.

DUELLMAN, W.E. and TRUEB, L. 1986. Biology of amphibians. McGraw-Hill, New York.

DUTTA, H. M.; ADHIKRI, S.; SINGH, N. K.; ROY, P. K. MUNSHI, J.S.D. Histopathological changes induced by Malathion in the liver of freshwater catafish Heteropneustes fossilis (Bloch). **Bulletin of Environmental Contamination and Toxicology**, v.51, p.895-900, 1993.

EMBRAPA - BRAZILIAN AGRICULTURAL RESEARCH COMPANY. **Pollution by pesticides, nutrients and suspended material in the rivers that form the Pantcloacal region of Mato Grosso**. ADM - Media Release Article, Embrapa Pantcloacal, Corumbà, MS, n.096. p.1-4, mar. 2006. 2004

FANTA, E.; RIOS, F. S.; ROMÂO, S.; VIANNA, A. C. C.; FREIBERGER, S. Histopathology of the fish *Corydoras paleatus* contaminated with sublethal levels of organophosphrus in water and contaminated with sublethal levels of organophosphorus in water and food. **Ecotoxicology and Environmental Safety**. V54. 2003.

FARIA, L, J da SILVA. **Evaluation of different sorbents in the solid phase extraction of pesticides in water**: development and validation of a methodology. Master's dissertation. Institute of Chemistry. UNICAMP. 2004.

FDA. Food and Drug Administration. **Certain others from new animal drugs:** formalin solution. Code of federal regulations. 21 CFR, Part, 529. 130. Washington, 1998. Available at: <www.fda.org>. Accessed on August 18, 2008.

FDA. Food and Drug Administration. **Requirements for investigational new animal drugs. Eastern fish health group and the American fisheries society fish health.** The FDA Workshop. Alabama. 1992. Available at: <www.fda.org>. Accessed on August 18, 2008.

FEDER, M. E., and BURGGREN, W.W. **Cutaneous gas exchange in vertebrates**: Design, patterns, control, and gill ventilation. J. Exp. Biol. 90. 1985.

FEIO, R.N.; *et al.* **Amphibians of the Rio Doce State Park (Minas Gerais).** Viçosa: UFV/Instituto Estadual de Florestas, 1998.32p.

FEPAM. **Industrial Liquid Effluents**: Pollutant Loads Released into the Water Bodies of Rio Grande do Sul. Porto Alegre: Henrique Luiz Roessler State Foundation for Environmental Protection, 2007. 145 p.

FREITAS, M.A.; SILVA, T.F.S. **Anfibios da Bahia**: Um guia de identificaçâo. Camaçari, 2004.

GEORGE, L. L. e CASTRO, R. R. L. **Histologia comparada**. 2. Ed. rev. e ampl. Sâo Paulo: Roca, 1998.

GRISOLIA, C. K. **Pesticides**: mutations, cancer and reproduction. Brasilia, DF: University of Brasilia, 392p. 2005.

HADDAD, C. F. B. **Identification key: anuran amphibians from the Jundiai slope of the Serra do Japi, state of São Paulo**. UNICAMP. Campinas - SP. 2005. Accessed on 01/03/2010, available at: http://www.biotaneotropica.org.br/v5n2/pt/abstract?identification-key+bn03005022005.

HADDAD, C.F.B. & R.P. BASTOS. 1997. Predation on the toad Bufo crucifer during reproduction (Anura: Bufonidae). **Amphibia-Reptilia** 18: 295-298.

HADDAD, C.F.B. Biodiversity of amphibians in the State of Sâo Paulo, pp.15-26. In: Castro, R.M.C. (ed.). **Biodiversity of the State of São Paulo, Brazil** (Vertebrates Series). 71p. FAPESP, Sâo Paulo. 1998.

HADDAD, C.F.B.; ABE, A.S. **Amphibians and Reptiles**. In: Workshop Atlantic Forest and Southern Fields. 1999. Available at: <http://www.bdt.org.br/workshop/mata.atlantica/BR/rp_anfib>. Accessed on: Feb. 22, 2007.

HADDAD, C.F.B.; POMBAL Jr., J.P. & R.P. BASTOS. **New species of Hylodes from the Atlantic Forest of Brazil** *(Amphibia: Leptodactylidae)*. Copeia, 1996(4):965-969.

HASCHEK, W. M.; ROUSSEAUX, C. G. **Handbook of toxicology pathology**. London. Academic Press, 1996. 127-151 p.

HEATH, A. G. **Water pollution and fish physiology**. Virginia: Lewis Publishers, 1995. 359p.

HEYER, W.R.; *et al.* **Measuring and monitoring biological diversity**: Standard methods for Amphibians. Washington: Smithsonian Institution Press, 1993.

HOFLING, E. et al. **Chordata**: manual for a practical course. Sao Paulo: Edusp, 1995.

IBAMA. http://www.ibama.gov.br. Accessed on November 11, 2009.

IBANEZ, M., PICÓ, Y., MÂNES, J. **Chromatogr.** A, 728. 1996. 325p.

IBGE, **Research Directorate, Population and Social Indicators Coordination, Municipal Basic Information Survey,** 2002.

IBGE. **Research Directorate, Population and Social Indicators Coordination, Municipal Basic Information Surveys.** 2005.

IBGE. **Research Directorate, Population and Social Indicators Coordination, Municipal Basic Information Surveys.** 2007.

JOHNSON, M. P.; SWETNAM, L. D. **Sprayer nozzles:** selection and calibration. Lexington: University of Kentucky, 1996. 6 p.

LANDRES, P. B. Ecological indicators: panacea or liability? In: MCKENZIE, D. H.; HYATT, D. E.; McDONALD, V. J. (eds.). **Ecological indicators.** New York: Elsevier Applied Science, 1992.

LANGIANO, V.C. **Toxicity of roundup and its effects on the neotropical fish prochilodus lineatus**. 2006. 62 p.Dissertation (Master's Degree) - State University of Londrina, Londrina.

LECH, J.J.; VODICNIK, M.J. Biotransformation. In: Rand, G.M., Petrocelli, S.R. (Eds.), Fundamentals of Aquatic Toxicology; **Methods and Applications**. 1985. p. 526 -557.

LÉVÊQUE, M. Environmental monitoring through bioindicators: a French-Brazilian cooperation in the study of foraminifera. **France-Flash Environment.** n. 9, p. 1-10, 1996.

LORENZI, T.F. **Manual de hematologia**. Sao Paulo, MDSI. 2006, 641p.

MACHADO, P. A. L. **Brazilian environmental law**. 11ª ed. Revised, updated and expanded. Sao Paulo - SP. 2003. 1064p.

MACHADO, R. **Presentation in a mini-course on sampling methods**, 4th Congress of Herpetology, Pirenópolis - GO, 2009.

MARTINEZ C.B.R.; CÓLUS, I. M.S. Biomarkers in neotropical fish for monitoring aquatic pollution in the Tibagi River basin. In: MOACYR, E.MEDRI; BIANCHINI,E.;SHIBATTA,O.A.;PIMENTA, J.A. **A bacia do Rio Tibagi**. Londrina: MC Grafica, 2002. p.551-577.

MARTINS, A. K. E. **Ipucas da Planicie do Araguaia, state of Tocantins: physical environment of occurrence, soils and land use.** 2004. 168p. Thesis (Doctorate in Forest Science) - Federal University of Viçosa, Viçosa, MG.

MCNEELY, J. A and SCHERR, S. J. **Ecoagriculture**: feeding the world and biodiversity. Sao Paulo: Senac Sao Paulo publishing house, 2009.

MEDEIROS, M. B. de. FILHO J. R de M.; WANDERLEY P. A. Agrotoxics in the Rural Environment: A Constant Threat to Agricultural Environments. In: II Meeting on the Environment and Environmental Education at UFPB. **Abstracts**. Joao Pessoa: PRAC/COPREX, 2008. v. 1. p. 78-79.

MEYER, A. *et al.*, Cancer mortality among agricultural workers from Serrana Region, State of Rio de Janeiro, **Brazil Environmental Research,** Basel, v. 93, n. 3, p 264-71, nov. 2003.

MILLER, G. T. (1931). **Environmental Science.** Technical review: Thomson Learning, 2007. ªOriginal Title: Environmental science "Translated from the Latin American edition". 2007.

MMA. MINISTRY OF THE ENVIRONMENT. Available at: http://www.mma.gov.br/sitio/index.php?ido=conteudo.monta&idEstrutura=143&idMenu=7242.

Accessed on February 24, 2010.

MONTENEGRO, A. T. **Science, history and memory**: methodological issues. In. ERTZOGUE, M. H.; PARENTE, T. G. (org.). *História e Sensibilidade*. Brasilia/DF: Paralelo 15, 2006.

MONTENEGRO, R.M.; FRANCO, M. **Pathologia**: processos gerais. 4. ed. Sâo Paulo: Editora. Atheneu, 1999. 320 p.

MURT, C. R.K., NAG, D. Human health impact of pesticides in the environment, in Chemistry, agriculture and the environment, M. L. Richardson, ed., **The Royal Society of Chemistry,** Cambridge. 1991. 491p.

NAVAL, *et al.*A reflection on the implementation and socio-environmental sustainability of the Luis Eduardo Magalhaes HPP - TO. **Brazilian Journal of Research and Development**. Vol. 10. N° 2. August - 2008.

NESKOVIC, N. K.; POLEKSIC1 V.; ELEZOVIC2 I.; KARAN, V.; BUDIMIR, M. Biochemical and histopathological effects of glyphosate on Carp, Cyprinus carpio L. Bull. **Environ. Contam. Toxicol**, v.56, p.295-302, 1996.

NOGA, E.J. Fish disease: diagnosis and treatment. Mosby: North Carolina State University, St. Louis: College of Veterinary Medicine, 1996. 367p.

PIMENTEL,D. Green revolution agriculture and chemical hazards. **Science of the total environment**, Netherlands, v. 188, n. 1, p. 86- 98, 1996.

POUGH, F. H.; HEISER, J. B.; MCFARLAND, W. N. **The Life of Vertebrates**. 2nd Edition. Atheneu Editora, Sao

Paulo, 1999. 798 p.

POUGH, F. H.; HEISER, J. B.; MCFARLAND, W. N. **The Life of Vertebrates**. 2nd Edition. Atheneu Editora, Sao Paulo, 1999. 798 p.

RABB, G. B. **Declining amphibian population. Species**, 1990. 33-34.

RACH, J.J.; HOWE, G.E.; SCHREIER, T.M. Safety of formalin treatments on warm and coolwater fish eggs. Aquaculture, Amsterdam, v. 149, p. 183-191. 1997.

RAMSDORF, Wanessa. Use of two species of Astyanax (Astyanax Sp B and A. Altiparanae) as bioindicators of a region contaminated by pesticides (Canguiri farm - UFPR) 2007. 127 f. Dissertation (Master's Degree in Genetics) - Federal University of Paranâ, Curitiba.

RAZ, G. A. Crustaceans and Polychaetes. In: DE LA LANZA, E. G.; HERNANDEZ, P. S.; CARBAJAL, P. J. L. (eds). **Organisms that Indicate Water Quality and Contamination (Bioindicators)**. Plaza y Valdes. Mexico. 633 p. 2000.

ROBERTS, R.J. **Pathologia de los peces**. Madrid: Ediciones Mundi-Prensa, 1981. p. 367.

RODRIGUES, E.L.; FANTA, E. Liver histopathology of the fish Brachydario rerio after acute exposure to sublethal levels of the organophosphate Dimetoato 500. **Revista Brasileira de Zoologia**. Curitiba, v. 15, p. 441-450, 1998.

RUEGG, E.F., PUGA, F.R.; SOUZA, M.C.M. de; ÙNGARO, M.T.S.; FERREIRA, M. da S.; YOKOMIZO, Y.; ALMEIDA, W.F. **Impacto dos agrotóxicos sobre o meio ambiente, a saùde e a sociedade**. Sao Paulo, ed. icone, 94 p, 1991.

SBH. Brazilian Society of Herpetology. Available at: http://www.sbherpetologia.org.br/checklist/checklist_brasil.asp. Accessed on: January 25, 2010, at 17:37 hours.

SILVA, A. F.; *et al.* Floristic composition and ecological groups of species in a section of Submontane Semideciduous Forest on the Sao Geraldo farm, Viçosa, MG. **Revista Arvore**, v.27, n. 3, p. 311319. 2004;

SILVANO, D. L.; COLLI, G. R.; DIXO, M. B. de O.; PIMENTA, B. V. S.; WIEDERHECKER, H. C. **Amphibians and Reptiles**. Department of Zoology. University of Brasilia. Brasilia - DF. 2003.

SILVANO, D. L; SEGALLA, M. V. Conservation of amphibians. **Megadiversidade,** Brasilia. v. 1. n. 1, p. 79-86, July 2005.

SOTHERTON, N. W. Landuse and the decline of farmland wildlife: an appraisal of the set-aside approach, **Biological,** n°, 1998.

SOUZA, R. T. Uso de Equipamentos de Proteçao Individual na Pulverizaçao de Videiras. Porto Alegre: EMBRAPA, 2006, p. 8 -10. Available at: < http: www.cnpuv.embrapa.br > Accessed on: October 13, 2007.

SOUZA, V.L.; LUNARDI, L. O.; VASQUES, L. H.; CASALETTI, L.; NAKAGHI, L. S. O.; URBINATI, E.C. Morphometric alterations in fepatcytes and ultrastructural distribution of liver glycogen in pacu (*Piaractus mesopotamicus Holmberg,* 1887), during food restriction and refeeding. **Brazilian Journal of Morphological Sciences,** v.18, n.1, p.15-20, 2001.

STEBBINS, R. C.; COHEN, N. W., **A Natural History of Amphibians**. Princeton University Press. New Jersey, 1995, 316p.

TAYLOR, G. R. **A ameaça ecològica.** ed. Brasileira rev. prefacada por Màrio G. Ferri; traduçao de Rodrigo Machado.

Sao Paulo: Verbo: Ed. da Universidade de Sao Paulo, 1978.

TERRA VIVA. **Fertilizers time to turn the tables**. Terra Viva magazine. ISSN 1983-1099. Year 2 - n°15. October 2009.

TOCANTINS. Atlas of Tocantins: Subsidies for territorial management planning. Planning Secretariat - SEPLAN. Superintendence of Planning and Central Management of Public Policies. Ecological-Economic Zoning Directorate - DZE. Organized by Ricardo Ribeiro Dias, Eduardo Quirino Pereira and Lindomar Ferreira dos Santos. 5 ed. rev. actual. Palmas. SEPLAN, 2008. 62p.

TOCANTINS. **Environmental impact assessment and environmental impact report for the Jalapao State Park visitor reception center**. Palmas: SEPLAN, 2003.

TOCANTINS. Government of the State of Tocantins. Tocantins State Development Company - CODETINS. **Javaés Project**. Lagoa Irrigation Subproject. Soil Study. Palmas: 1997. 104 p.

TOCANTINS. State Law No. 1.560 of April 5, 2005, which establishes the State System of Nature Conservation Units - **SEUC**. Palmas - TO. Available at: http://srhma.to.gov.br/. Accessed on November 12, 2009.

TOCANTINS. State Law No. 1.560 of April 5, 2005, which establishes the State System of Nature Conservation Units - SEUC. Palmas - TO. Available at: http:// srhma.to.gov.br/. Accessed on August 3, 2009.

TOCANTINS. Secretary of Water Resources and Environment of the State of Tocantins. Available at: http://recursoshidricos.to.gov.br. Accessed on January 20, 2009.

TOCANTINS. Nature Institute of Tocantins. Available at: http://naturatins.to.gov.br/. Accessed on November 5, 2008.

TOWNSEND, C. R. **Fundamentals of ecology**. Translated by Gilson Rudinei Pires Moreira... (*et al.*). - 2. ed. - Porto Alegre: Artmed, 2006. 592 p.; 25 cm.

TRAPÉ, A. Z. The case of pesticides. In. ROCHA, L. E. *et al.* (Org). **Is this people's work?** Vida, doença e trabalho no Brasil, Petrópolis: ed. Vozes, 1993.

VAN DYK, J.C. **Fish histopathology as a monitoring tool for aquatic health**: a preliminary investigation. 2005. 47f. Thesis (PhD) - University of Johannesburg, Johannesburg, South Africa.

VIAL, J. L. **Declining amphibian populations task force. Species.** 1991. 47-48.

VIANA, V. M.; TABANEZ, A. A. J. Restoration and management of forest fragments. In: Congresso Nacional sobre essências nativas, 2, 2992, Campos do Jordao. **Proceedings...** Campos do Jordao: Instituto Florestal/Secretaria do Meio Ambiente, 1992.

WWF. **Analysis of the environmental impacts of farming activities in the Cerrado and their interrelationship with water resources in the Pantcloacal region**. Consultancy by Andréa Aguiar Azevedo. ECO/UnB. 2001.

ZAMBRONE, F. A. D. **SINTOX**: Toxicological Information System. Rio de Janeiro: Fiocruz, 2002, p.2-43. Available at: <http://www.fiocruz.br.cict.oque.estrut.dect/sintox.kitbrasil.html> Accessed on January 2, 2009.

ZUCKER, E. Hazard evaluation division. Standard evaluation procedure. **Acute toxicity test for freshwater fish**. USEPA publication 540/9-85-006, Washington, D.C., 1985. Available at: <www.epa.gov>. Accessed on: March 18, 2009.

CHAPTER II QUANTITATIVE AND QUALITATIVE SURVEY OF AGROTOXIC PRODUCERS AND COMMERCIAL ESTABLISHMENTS

The use of chemical substances as a way of controlling or eliminating problems arising from attacks by pests and diseases on cultivated plants and livestock can be traced back to classical antiquity. One of the difficulties in protecting natural ecosystems is that there are differences in the perception of values and importance between individuals from different cultures or socio-economic groups who play different social roles in these environments. Various factors affect the values adopted by farmers, since value is relative to the needs of each individual. Based on these assumptions, the aim of this chapter was to understand how farmers think, based on a group of interviewees, and what ecological conceptions they have produced based on a common practice. Nine farmers responsible for irrigation projects located at points 2 and 3 of the study were interviewed. Of those interviewed, one did not answer which pesticides were used in the hydro-agricultural project, while the others said they couldn't remember all the ones they used, but they did give the name of one they used. They usually gave the names of herbicides and insecticides, as they said they were most commonly used. The list given to them by the commercial establishment they were looking for listed some of the most commonly used pesticides, while the nine interviewees said that they only remembered using 23 pesticides, and that for the project alone, approximately 70 different active ingredients are sold every six months, i.e. for a specific type of crop, such as irrigated rice, since an even greater number of compounds are sold for soya and other crops, since these require different active ingredients from those used for rice. During the interviews, the dependence on pesticides for good crop results was noted. Based on the results obtained from the interviews, a picture emerged of vagueness and contradictions between verbal discourse and actual practice in agricultural projects, such as the use of some agrochemicals that can cause damage to the aquatic environment and the organisms that depend on it. The agrochemicals used by farmers are permitted under Brazilian law, but some have been banned due to their carcinogenic and mutagenic properties. As a result of the accumulation that can occur in the living tissues of organisms that are exposed to pesticides, the long-term risk posed by the constant use of pesticides in this region, which has a rich biodiversity, is a cause for concern. In this way, some essential precautions need to be intensified for the region so that it is not only unproductive but also negatively impacted.

Keywords: Interviews, pesticides, irrigated rice, Lagoa da Confusao.

1 - INTRODUCTION

The use of chemical substances as a way of controlling or eliminating problems arising from attacks by pests and diseases on cultivated plants and livestock can be traced back to ancient times (ALVES FILHO, 2002). The fight against pests and diseases that affect agriculture and forestry, following the development of synthetic products and the widespread dissemination of this technology in all countries, has always led to a debate about the real harms and benefits of pesticides (ALVES FILHO, 2002).

A pesticide is a chemical or biological product that is used in areas of production, storage and processing of agricultural products, pastures, protection of forests and other ecosystems, as well as urban, water and industrial environments. Its purpose is to alter the composition of flora or fauna in order to preserve them from the damaging action of harmful living beings (Law No. 7.802, of 1107/1989).

Pesticides, especially those of organic origin, have been used since 40 years after Christ. The introduction of organosynthetic pesticides in Brazil began in 1943, with the first samples of DDT (SPADOTTO, 2006). In the existing agricultural model, pesticides are considered indispensable and the main chemical pollutants that spread across the planet. The active ingredients in pesticides are used in a variety of urban and rural situations (GRISOLIA, 2005).

The indiscriminate application of pesticides affects both human health and natural systems and it is estimated that they are responsible for more than 20,000 unintentional deaths a year, the majority of which occur in the Third World, where around 25 million agricultural workers are poisoned (COUTO, 1998).

According to MONTENEGRO (2006) "the work of producing documents from interviews is an activity that, associated with its technical specificities (interview script, place and time of recording, transcription and editing technique, letter of release) and methodological ones (above all related to the understanding of memory and the possible insertion of these records in historiographical projects), establishes the production of a documentary source".

In this type of work, it is customary to use concepts of environmental perception, where it is the operation that exposes the logic of the language that organizes the expressive signs of the uses and habits of a place. It is an explanation of the image of a place, conveyed in the signs that a community builds around itself, which can be revealed by reading the discursive production of a community (FERRARA, 1993).

From the 1930s to the present day, dams have been built on almost every river in the world. There are several examples in Brazil, such as: the Paranâ River was dammed at Itaipú, generating more than 10,000 MW of installed capacity, at the cost of flooding spectacular landscapes (ALIER, 2007). In the state of Tocantins, several dams can also be identified for the purpose of building hydroelectric plants, some of which have already been built and others planned, with around 14 in operation, generating 1,403.35 MW (SEPLAN, 2008).

The largest consumption of *water* in Brazil is in agriculture, with around 70%, of which the majority is destined for irrigation, which comes from the conservation of rivers, which are constantly being degraded by the deforestation of riparian forests and the dumping of pollutants on a scale greater than the maximum absorption limit of the rivers, altering their dynamics. The pesticides used negatively affect all ecosystems, with aquatic pollution being one of the most worrying due to its scarcity (COSTA, 2004).

Large irrigation projects include dams, lakes, pumping units, canals and pipes, a water distribution system for the plots and a drainage system. In Brazil, one of the problems involving the dispute over the use of water is irrigation, where its expansion in some basins, following the implementation of several irrigation projects without prior quantification of the volume of water available, has led to a lack of water for areas located downstream, as well as contamination of the irrigator during irrigation, the community near the irrigated area and the user of irrigated products (BERNARDO, 1992).

The aim was to carry out a quantitative and qualitative survey of information on pesticides used by producers, as well as those marketed by pesticide sales outlets in Lagoa da Confusao - TO.

2 - MATERIALS AND METHODS

This study was carried out after obtaining permission to carry out the interviews, following the protocols issued by the Ethics Committee of the Lutheran University of Palmas - CEULP- ULBRA, as shown in Annex I.

2.1 - Location of interview points

The municipality of Lagoa da Confusao, with an area of 10,565 km^2 , is part of the Araguaia National Park, where the sub-basin of the Urubu River is located. It is one of the eight sub-basins of the Formoso River, which covers some 2,702 km^2 and its main course runs for 130 km to its mouth. The basin has an elongated transverse shape, narrowing towards the mouth of the Formoso River. The main tributaries of the Urubu River, on the left bank, are: Ribeirao Barreiro, Rio Tioribero and Ribeirao das Pedras. On the right bank, the Ribeirao Cangirana, in its middle course, stands out as the main tributary (TOCANTINS, 2007). The study was carried out between the geographical coordinates W 49° 41' 17.54", S 10° 50' 49.32" and W 49° 2' 43.99", S 10° 38' 13.84", in three areas of the Urubu River, where two were necessarily marked by the presence of agricultural enterprises and the other by the presence of native vegetation cover, free from agricultural projects.

Traveling along the course of the Urubu River was necessary to recognize the area of the Urubu River and identify and demarcate the study points. Figure 1 depicts the reality of point 1, which had no agricultural projects in its vicinity.

Figura 1: Recognition and identification of the study points, this being point 1 demarcated for study, Cristalândia - TO, 2008. Source: ROCHA, 2008.

The distinction between the points chosen along the Urubu River for this research was made by checking satellite images, topographic maps and field visits to confirm the study area, as well as by identifying the location of the points of interest through guidance provided by local residents. The map shows the delimited research area (Figure 2).

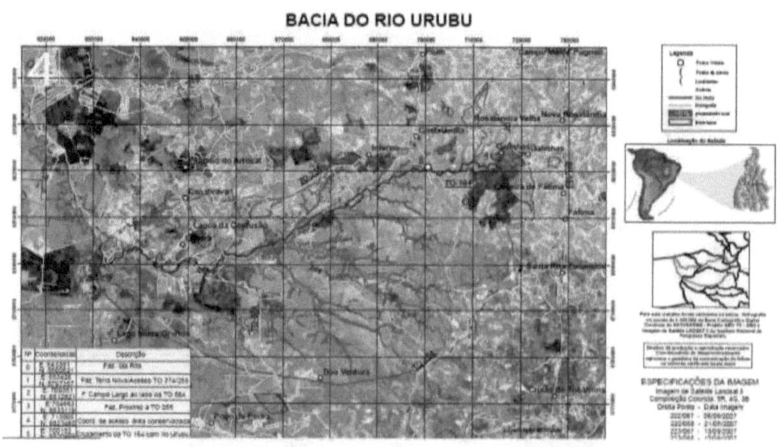

Figura 2: Delimitation of the Urubu river sub-basin, Lagoa da Confusao and Cristalândia - TO, 2009. Source: NATURATINS, 2008.

The areas of the farms were all located in the municipality of Lagoa da Confusao, where there is the highest concentration of agricultural projects, which in a way motivated the choice of location for the study. A total of 15 people were interviewed, among them nine owners of agricultural projects who are symbolized in Figure 3 by small green circles. Some of the numbers are not represented in the figure because they coincide with other producers, due to the fact that the agricultural project areas are located very close to each other, so we opted to take the coordinates of a single interviewee when this happened. However, the interviews conducted were all concentrated near the Urubu River, in the part that comprises the municipality of Lagoa da Con fusa o - TO (Figure 3).

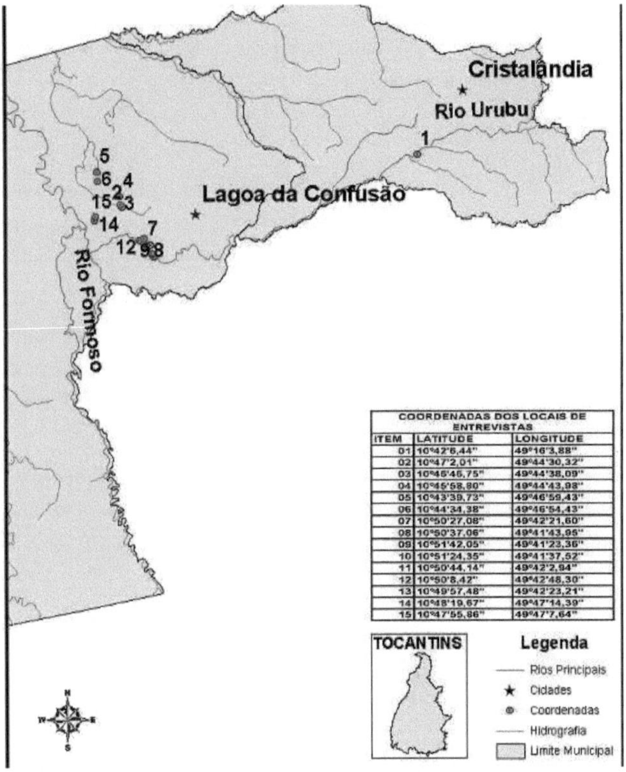

COORDENADAS DOS LOCAIS DE ENTREVISTAS		
ITEM	LATITUDE	LONGITUDE
01	10°42'6,44"	49°16'3,88"
02	10°47'2,01"	49°44'30,32"
03	10°46'46,75"	49°44'38,09"
04	10°45'58,80"	49°44'43,98"
05	10°43'39,73"	49°46'59,43"
06	10°44'34,38"	49°46'54,43"
07	10°50'27,08"	49°42'21,60"
08	10°50'37,06"	49°41'43,95"
09	10°51'42,05"	49°41'23,36"
10	10°51'24,35"	49°41'37,52"
11	10°50'44,14"	49°42'2,94"
12	10°50'8,42"	49°42'48,30"
13	10°49'57,48"	49°42'23,21"
14	10°48'19,67"	49°47'14,39"
15	10°47'55,86"	49°47'7,64"

Figura 3: Cartography of the Urubu River, Lagoa da Confusao and Cristalândia - TO, where animals of the species *Leptodactylus ocellatus* were captured, 2009.

Source: BASTOS, 2010.

The number of samples is due to the fact that the area around the river, where the study points were delimited, has few producers, but with large planted areas. Sometimes there was more than one visit to the same place and interviews were carried out at the same point a second time, since the areas are leased and it can happen that a certain producer only grows rice in one period and in the second semester another producer uses the same region to plant other crops, such as beans, soybeans, for example.

In fact, the whole process of identifying the properties began in the second half of 2008, with *on-site* visits accompanied by a PhD student who was already working in the region surveying cerrado amphibians. The interviews began in March 2009, and every time we visited the properties we also interviewed the farmers who were located close to the two capture points and who had not yet been interviewed.

During 2009, the team, always made up of more than two people, visited the study points, i.e. places where agricultural crops were planted, interviewing the producers, who were identified by numbers in order to preserve their anonymity. To reach the potential interviewees, visits were made to farms located near the Urubu River, and the project was visited to find the person responsible for the first conversation. Once they had agreed to take part in the research, all the material used was prepared, and if the time was suitable for the participant, the work began right there, otherwise they returned to the site at an opportune time scheduled by the participant.

The methodology for the interviews used in this research was qualitative, which is interesting because "it is based on a strategy based on data collected in social or interpersonal interactions, analyzed from the meanings that subjects and/or the researcher attribute to the fact". In this way, the researcher sets out to participate, understand and interpret the information (CAMPOS, 2004).

This study prioritized field research techniques in environmental perception based on three types of approaches: observing, asking and listening, characteristic of the conceptual triangle model (WHYTE, 1977). At the time of the interviews, the interviewees were asked questions which required a great deal of caution on the part of the interviewer, avoiding inducing answers, thus obtaining a result that was closer to reality.

The methodology used to carry out the research was the oral history of agricultural producers, which is a modern resource widely used to compile documents, archives and studies of the social experience of people and groups (MEIHY, 2002). Oral history uses interviews with people previously selected on the basis of a script of questions relating to the subject, as well as recording their conversations.

There are two types of question: open-ended and closed-ended. Open-ended questions allow the respondent to construct the answer in their own words, thus allowing freedom of expression. Closed-answer questions are those in which the respondent simply selects the option (from among those presented) that best suits their opinion. It is also common for questions of both types to appear in the same script, which is considered to be mixed (CAMPOS, 2004). The interview script used open and closed questions, characterized as a mixed or semi-structured interview.

The questionnaires were analyzed using basic statistics and computer graphics. The data was processed using spreadsheets with a square matrix format, with the cases and variables arranged in rows and columns (ARANGO, 2005).

The interviews served as a basis for finding out information about the agricultural projects and the farmers, such as: the socio-economic profile of the interviewees; the job/activity involved; data on the property; information on how and when they planted and on the use of pesticides.

3 - RESULTS AND DISCUSSIONS

Below are the various opinions expressed by agricultural producers in the Lagoa da Confusao region, regarding the techniques adopted for cultivation, productivity, the use of pesticides, their advantages and disadvantages, as well as social information, among others presented in the interview form, see Annex I.

3.1 - Socio-economic profile of interviewees

One of the interviewees didn't want to take part when there was a technical fault with the recording equipment, which caused the interview to be interrupted, leaving the participant free to give up. The interviewee was a man, aged around 60, who had a small plantation, which, according to him, was only for his own consumption. And as a way of justifying not taking part in the survey, he said "as you've already interviewed many people out there, just copy what they've said... it's all the same for everyone here in Lagoa", in other words, for him, everyone in the Lagoa da Confusao region grew the same products and used the same agrochemicals, and he didn't think his participation was important.

During the interviews, which took place in March, July, September and October 2009, several plantation areas were visited and the following properties were selected for the interviews: Fazenda Lago Verde, Formosinha, Fazenda Santo Antônio, Fazenda Cheguei, Fazenda Santa Rita, Fazenda Fazendinha and Fazenda Luiza. On these properties we found producers of irrigated rice or soybeans, beans and watermelons, developing their hydro-agricultural projects on leased areas. Some of these producers remained in the leased field for a period of the year and in the second half of the year they left for other regions of the state of Tocantins to work on other land (either agriculture or livestock), but most of them are dedicated exclusively to the hydro-agricultural project, almost all of them attracted to the region by the abundance of water and the fertility of the soils, considered to be concretionary, comprising soils ranging from well-drained to strongly drained (TOCANTINS, 2008).

A total of 9 (nine) men were interviewed, aged between 20 and 59, two of whom had a 6th grade education, two had a high school education, two had completed high school, two were agricultural technicians and one had completed higher education, making up a total of 88.88% who had only a high school education and 11.11% who had higher education, as shown in Figure 4. These figures may portray the reality of the difficulty shown by most of them in gaining knowledge of other techniques that offer less impact.

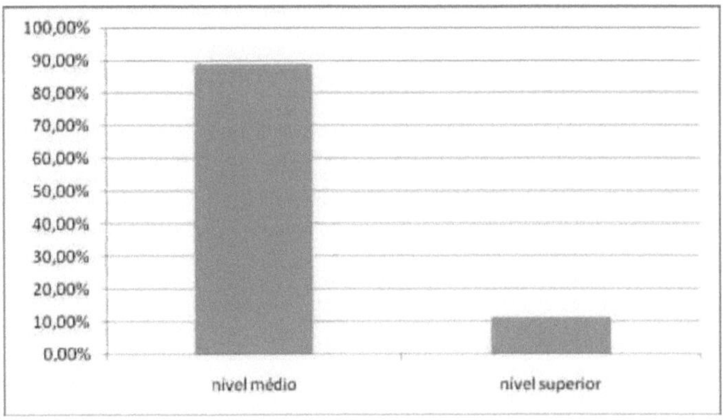

Figure 4: Educational level of the interviewees, Lagoa da Confusào - TO, 2009.

The women who were in the cultivation region were hired by the managers of the leased farm or by the producers who leased the land. They were responsible for routine household activities (cooking, washing clothes and taking care of the living environment shared by the workers hired for the project).

Most of the families come from other states, mainly the south of the country. The difference in age between the interviewees is due to the fact that many of the children grew up learning their father's trade, remaining there and developing their agricultural activities. This was evident during a visit the research team made, where they were greeted by a man who was harrowing the land. When asked who was responsible for the property, he said "it's my son Joào" (fictitious name, in order to preserve anonymity), who later answered the interview, a young man of 21.

The educational level of the interviewees, considered to be low, is mainly due to the fact that the children choose to follow their parents' profession, so most of them remain full time in the cultivated area and are involved in practically every stage of the crop, such as: preparing the land, applying agrochemicals, cultivating the selected crops, harvesting and transportation. Of the nine producers interviewed, only one of them outsources the land preparation and transportation stages.

Of the producers interviewed who farmed in the Lagoa da Confusao - TO region in 2009, 66.67% have the hydro-agricultural project as their only source of income and 33.33% carry out other types of activity, such as livestock farming, reselling pesticides and transporting water during the summer, according to Figure 5.

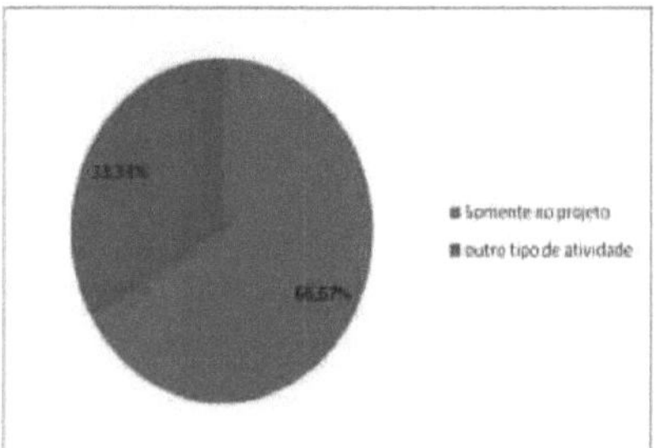

Figura 5: Producers' source of income, Lagoa da Confusao - TO, 2009.

Rural women are deeply aware of society's dependence on a clean and abundant environment and still protest when the scarcity and contamination of water, air and soil threaten their families' survival, but the view still remains that it is their responsibility to look after the house, kitchen and family. It is up to the children to adopt their father's work practices, and with this in mind they are taken to the countryside as children to begin their first lessons (AGARWAL, 1992).

The main source of income for farming families is based on agriculture; some other activity, such as livestock, for example, is used as a store of value (savings) or to produce food for the family (milk) (DOURADO, 1999). However, approximately 67% of producers in the study region have the project as their only source of income, even if they don't farm on the same leased area. Another 33% have an alternative activity, as they prefer or practice planting at one time of the year, either with irrigated rice or soybeans, and at the other time of the year they tend to work with cattle for those who have higher financial conditions and the others work in various activities (transporting water by truck), these are the ones who have small cultivated areas and in order to supplement their income.

3.2 - Property data, Lagoa da Confusao - TO, 2009

The knowledge that the tenants of the hydro-agricultural project had about the property, where they were present for at least half the year, was very limited, as the interviews revealed.

With regard to the farm having a legal reserve area, only one of the interviewees was able to inform of the existence of a "compensation area in Cantâo", but was unaware of the size of the area.

Of the producers interviewed, 44.44% said that there was no degraded area on the leased farm, while 44.44% were unaware of its existence and only 11.11% said that they knew of some kind of degradation on the property where the hydroagricultural project was being developed, stating that "I have observed the death of some species of trees, but I believe this is due to the change in soil, as native species are not adapted to the high level of water in the soil", as can be seen in Figure 6.

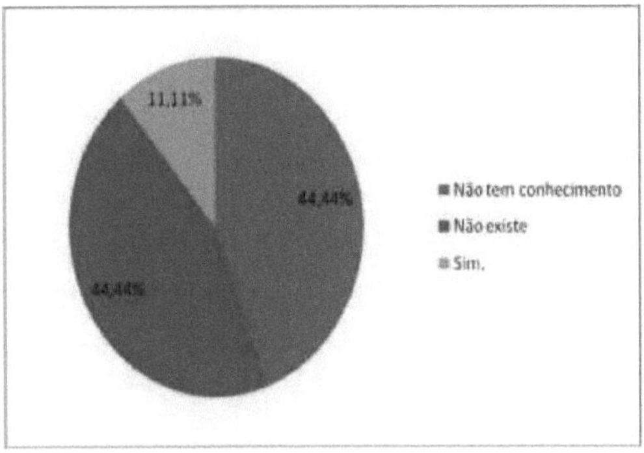

Figura 6: Existence of degraded areas, Lagoa da Confusâo - TO, 2009.

During the time spent visiting farms in the Lagoa da Confusâo - TO region, it was observed that there were some areas close to the properties where the research was carried out that were beginning to be degraded. Areas of native vegetation have been replaced by hydro-agricultural projects. Some signs of the removal of natural forest fragments were seen in the places visited, known as "ipucas", where this type of ecotonal vegetation existed and where the soil was unproductive, even though it had undergone corrective processes and was in the shape of a doline (MARTINS, 2002).

What has been clearly observed in that region is that land is really rare, which is why it has attracted so many producers from distant regions in search of fertile and productive land. Extensive areas have been deforested in order to cultivate them, leading some landowners to acquire farms, with the intention of increasing the size of their planted areas, as observed on farms located along a stretch of the Formoso River, a river that also has large hydro-agricultural projects.

Along the Urubu River, at certain points there is no riparian vegetation, which shows the presence of anthropic activities. In these places there were irrigation channels for agricultural projects and constructions such as houses or dams. The dam was no longer working due to a design error (according to residents) and was deactivated, but concrete blocks remained on the bank and in the riverbed.

The deforestation that takes place to build hydroelectric dams has far-reaching implications, as it results in the loss of biodiversity, often that which is still unknown, the depletion of nutrients in the soil due to leaching, the replacement of several hectares of forest biomass with livestock and, finally, the loss of carbon deposits and water evaporation capacity, putting not only regional and global ecological systems at risk, but also human survival on the planet (ALIER, 2007).

It is important to compare what the Lagoa da Confusâo region was like with some of the practices adopted in remote times.

The most common characteristic of primitive peoples is their respect for the land, which is considered to be the source of life. This was the principle adopted by Americans in ancient times. The land was a living thing, treated with affection, a place from which they could find cures for their ailments, or the God who answered their requests for rain, using dance to invoke him. For a long time, the Indians lived in this way, as the sole owners of the land, which remained intact and the scars left behind were insignificant, as the land was cultivated with skill and respect (UDALL, 1963).

Following UDALL's line of thought (1963), and applying it to the present day, we can see that environmental conservation is precisely what the Indians of old did, treating human beings as part of nature and not something outside of it.

In this sense, returning to the analysis made during the visits and in conversations with many other people who were encountered during the time of data collection, it is possible to identify that the intention of some producers who make use of land in that region is that, when they find a perfect place, which they were looking for, and which does not exist in some regions of Brazil, they settle there, and manage to produce sufficient quantities to export the production and make a good profit. In this way, they are known in the region as "big producers".

These producers, having left land that was unproductive in other regions of the country, or in search of land that would produce more, may well remain in the Lagoa da Confusâo region for as long as it is opportune. This intention is manifested by the fact that they lease land and don't buy it, thus making it quick for them to leave for any other region. This attitude is a concern for local people or those who have some appreciation for the place, because if local environmental degradation is promoted in any way, they will undoubtedly be the ones most affected.

The great devastation of natural resources began with mountain explorers and hunters and reached a series of climaxes in the last decades of the nineteenth century. The first phase involved only the destruction caused by primitive tools, but the second was linked to the machines of the industrial revolution. These made it possible to harvest crops on a large scale and also to destroy the soil on a large scale (UDALL, 1963).

The hydro-agricultural projects in the study region are considered to be large-scale, and heavy machinery is used to achieve the desired results. The people in charge of the projects have practically all the equipment on each leased farm, with a few being rented out when necessary, such as the planes used to deliver pesticides.

Farmers in the region chose to use two planting techniques, conventional and no-till. The areas used for production varied according to individual financial conditions and loans obtained through financing. According to the information they provided, most of them preferred conventional planting, because they were used to it and because it offered greater productivity, even for a smaller area (176 ha) it was possible to obtain a greater quantity of product (14,000 bags) compared to the other area, which was 250 ha and produced 13,750 bags, as can be seen in Table 1.

PRODUCTIVITY (Sacas)	**AREA (ha)**	**TECHNICAL (c/d)**
0	390	C
14.000	176	C
13.500	900	C
67.500	900	C
37.800	315	D
0	710	C
13.750	250	D
4.900	50	c
445.500	3.300	c

Table 1: Production data from the units visited, Lagoa da Confusao - TO, where C = Conventional planting technique; D = Direct, monitored during 2009.

Irrigated rice is planted in the Lagoa da Confusao region in the first half of the year, which is flooded due to the rainy season, usually from October to April. Of the total number of people interviewed for this study, 33.33% worked only with irrigated rice, while another 33.33% worked from May to September with other types of crops, ranging from beans to soybeans and even transgenic soybeans. A further 33.33% only grow beans from May to November and watermelons from April to July, and do not carry out any other agricultural projects in the region, as shown in Figure 7.

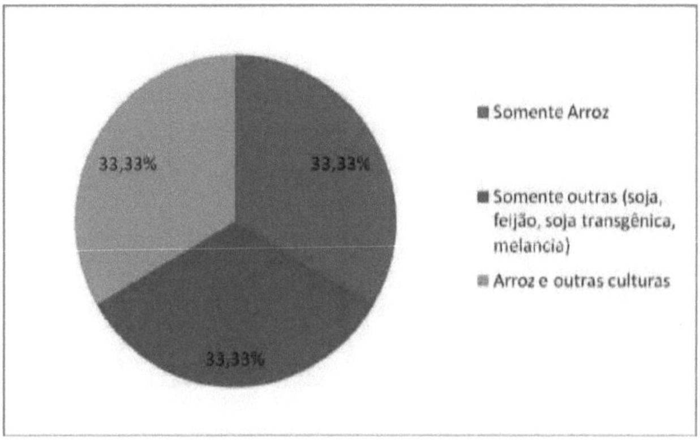

Figure 7: Type of cultivation practiced, Lagoa da Confusao - TO, 2009.

3.3 - Data related to the use of pesticides

One of the interviewees didn't answer which pesticides were used in the hydro-agricultural project, while the others said they didn't remember all the pesticides they used, but they did give the name of one they used. Generally, they gave the names of herbicides and insecticides, as they said they were most commonly used.

The pesticides they listed were: "*Gool, Hortene, Glifosato, glòria, acefato, Basugran, cobra, Fisológico, conecti, ZAP, Tamaron, Fosfato, Targa, Cloribrao, Bravonil, Folicul, Folegld, pivot, hortene, Decis, Karate, Tamaron, Certero*". Based on the list provided by the commercial establishment, the list of names presented was low, as nine interviewees reported a list of 23 names. However, for the project alone, approximately 70 different compounds are sold every six months, i.e. for a specific type of crop, such as irrigated rice, as an even greater number of compounds are sold for soy and other crops, as these require different active ingredients from those used for rice.

The pesticides were applied throughout the duration of the crop, starting after the harvest of the previous crop, to clean the land, which they call "desiccant", in order to plant the crop. 20 days after planting, the applications were resumed, which were carried out up to 55 days before the harvest, varying the date of the applications and the dosage of the pesticides according to the type of product applied, reaching a number of 8 applications per crop.

As for the destination of the pesticide containers, all the producers said they returned them, with 77.78% saying they returned them to the receiving station, which is the same place where they bought the pesticides, 11.11% said they had companies that went around collecting the containers from customers with their own vehicles, and 11.11% didn't say how they returned them (Figure 8).

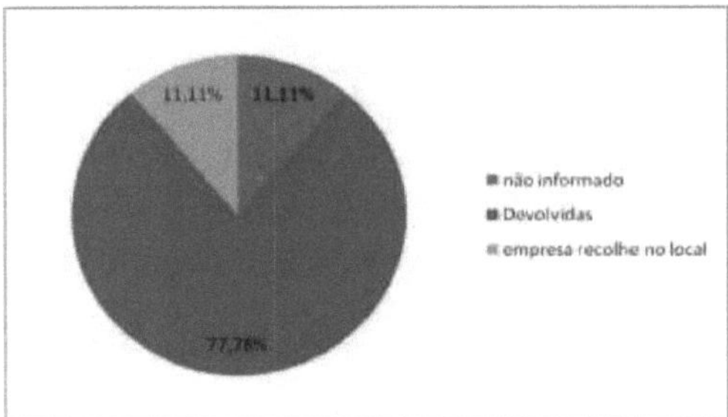

Figure 8: Return of pesticide packaging, Lagoa da Confusao - TO, 2009.

As for the method of application, 55.56% of producers used ground application of pesticides, but those producers who used to plant in both seasons used aerial application, as there was no way of making ground application, and it was only possible to do so by plane. There were still 22.22% who used both application techniques, 11.11% used only aerial application and 11.11% did not state which form of application they used, as shown in Figure 9.

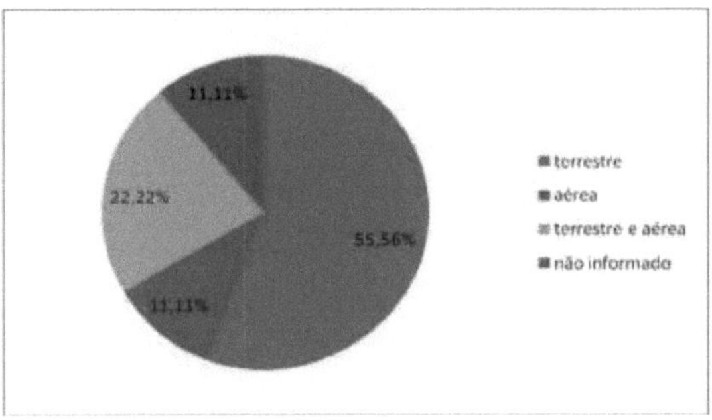

Figure 9: How pesticides are applied, Lagoa da Confusao - TO, 2009.

3.4 - Technical guidance on the use of pesticides

With regard to the use of Personal Protective Equipment (PPE), all the interviewees said they used it, as did the entire team working on the project, including those who applied it from the air, as was seen in one of the statements recorded: "*Use it. Anyone who works with poison has to use it, right? 90% of it is me who applies it.*" It is clear that they are aware of the need to use it, as well as its importance. However, after an interview during the reconnaissance visit to the project area, it was found that one of the employees who applied the pesticide using the uniport was not wearing PPE, was barefoot, wearing unsuitable clothes (shorts and a T-shirt) and no mask. Another employee, who was preparing the product in a truck specifically designed for this function, was wearing PPE (a lab coat, but the mask was not worn correctly, it was simply attached to his neck).

It was common to find working people, including those responsible for the project, walking around throughout the growing season, specifically during the pesticide application season, both on the ground and from the air, wearing inappropriate clothing (shorts and open shoes). Some of the people in charge of the project's support houses, mainly women, complained that during pesticide application the smell was very strong, as the wind helped to carry the sprayed product to the houses where they lived.

Before preparing pesticides, it is recommended that leaflets and labels be read. Of those interviewed, 88.89% said that they read the labels, one of them, although he did, said that he found it difficult to read the leaflet due to the size of the font, the person responsible for the larger area project said that the farm has two agronomists, who are obliged to pass on all the necessary information to those responsible for the application and only 11.11% said that they only read it sometimes, as shown in Figure 10.

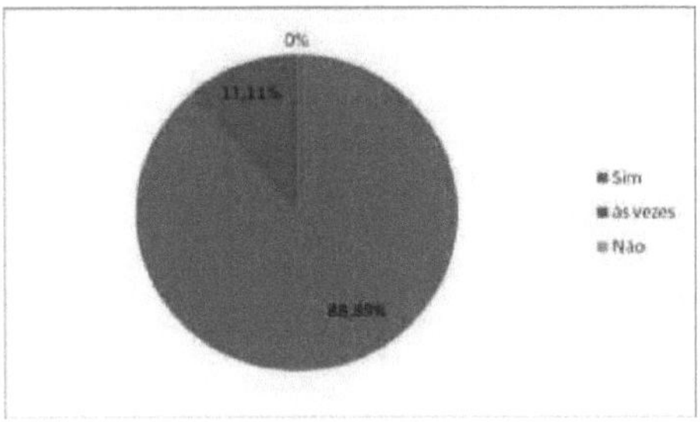

Figura 10: Reading pesticide labels, Lagoa da Confusao - TO, 2009.

Those projects that didn't have an agricultural technician available full-time had one who provided assistance from time to time, with visits ranging from 3 to 5 days, or whenever he was called. They also reported that the state's environmental inspection bodies (NATURATINS) and IBAMA visited all the projects constantly.

During the application of the pesticides, they said that there had never been any record of workers suffering any symptoms as a result of using the products. It happened on an agricultural project, as was actually posted on the property, how many days before the accident, with death, despite the fact that the pesticide was being applied, was not the cause but a traffic accident involving the machinery used by the workers.

When the producers were asked about the damage that pesticides can cause, they all said that there were harmful consequences. However, they use them so as not to lose their production, even though they are aware of the environmental risks caused by their use. They even reported that the use of pesticides does not bring any environmental benefits, which can be exemplified by one of them saying "everything that is chemical is not good for the environment".

According to the data collected, the producers said that they use pesticides to protect their crops and combat pests. They believe that they are more efficient and guarantee production, and unanimously state that they receive technical guidance on how to handle pesticides.

3.5 - Information on water used, Lagoa da Confusao - TO, 2009

The water used in the hydro-agricultural project for irrigation comes from the Urubu River, while the water they use for consumption comes from an artesian well, with the distance from the Urubu River to the well varying from 100 meters to a maximum of 4 kilometers.

The pumping system is on for 21 hours a day and is only switched off during the "green hour", which is the peak time between 6 and 9 p.m. During the rainy season, the time the pumping system is on is shorter, depending on the amount of water that is present in the plantation, thus reducing the consumption of water from the Urubu River.

The use of agrochemicals in the project area began after the harvest of the previous crop, applying *"desiccant to clear the land, limestone before planting, and immediately after planting, they apply fungicide and leaf fertilizer"*, according to one of the people responsible for the project. During the interviews, it was possible to identify that some type of agrochemical was applied throughout the project, but the most commonly used were insecticides (two or three times) and herbicides, which were applied more than once during planting, for example, Bim, "*fertilizer to avoid spots on the grains*", in relation to the watermelon crop, "sometimes *yes, sometimes no, because with the watermelon crop we look more at the plant and the weather and then we follow up and alternate the product as well".* The interval from one application to the next varied between the interviewees from one week to 15 days, depending on the cycle of the pesticide, however, if it was applied and there was rain, it should be applied immediately after the rain, in other words, according to the interviewees, in any case a certain product should be applied more than once, which varies according to the cycle of the product, the weather and the infestation of pests or insects, which are the main threats to the production process portrayed by them.

The main advantages identified by the producers in using pesticides are that they increase production and make it easier to control infestations of pests/weeds and insects, as can be seen in the words of some of the interviewees: interviewee 5 - *"It's easier to control because if you don't use pesticides nowadays you won't be able to harvest anything!";* interviewee 6 - *"Because nowadays there are infestations of fungi, pests, insects, so you have to control them if you don't, you can't produce"; interviewee 7 - "If you think about soybeans, there are a lot of severe caterpillar attacks here, but rice doesn't have a problem";* interviewee 8 - *"You'll have a crop clean of weeds, insects, and your productivity will be higher."* The others reported no advantages.

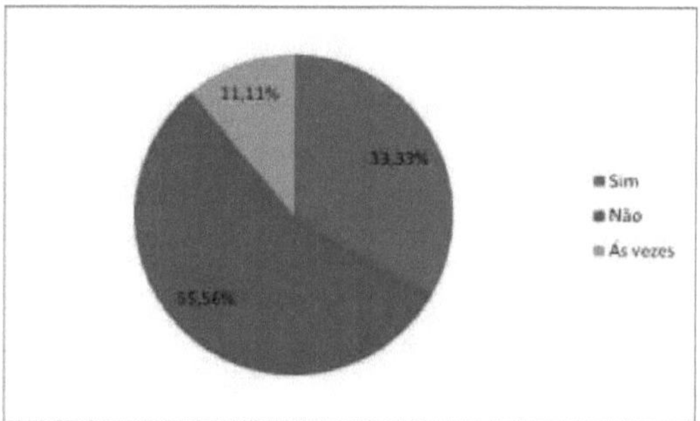

Figura 11: Disadvantages of using pesticides, Lagoa da Confusao - TO, 2009.

Figure 11 shows data related to the disadvantages that can be caused by the application of agrochemicals. Producers were asked if the products used could cause damage to the environment. 55.56% of those interviewed said that they do not cause damage, while 33.33% said that they cause some damage, as can be seen during the interviews, for example, Interviewee 2 - *"Foliage fertilizer, it doesn't bring results, no effect, of course"; Interviewee 5 - "It can be harmful to the environment if you use it outside the dosage, in excess, or throw it in areas where it can't";* according to him, it can't be used near the house, river forest. Interviewee 9 - *"Everyone talks to us. Land application can cause a risk of water contamination, that's something you can see, like people often take water straight from the lake, you can't do that. Now we have a tank to manage the water to avoid this, to wash the birds, the machinery, we have a cement base, with a seal, which directs the rest of the product that is thrown onto the crop."* Somehow he was trying to adapt his project in order to minimize the existing impacts; in fact, this was practically the only project we visited that was trying to follow environmental legislation for sustainable management. The project has a structure for storing pesticides, fire safety, animal entrances, as well as a suitable place for washing and storing empty containers, as shown in Figure 12.

All the producers said that it was impossible to produce, "*No, especially for soya because of the climate, it's very hot and the humidity is very high*", even more so on a large scale without the use of agrochemicals. One of them even commented that he knew of alternative practices, such as organic fertilization, "*like some people use to do in the South, but it wouldn't work here, where the plantations are large*". Another said that he had opted for another, less aggressive *method*: "*We're trying no-till, right, but even so... it's reduced a lot.*" In short, in order to summarize the thinking of the producers interviewed, one could use the following phrase from one of them "*Totally unfeasible, I'm not going to produce anything at all*", which reflects the unanimous thinking.

Figura 12: Storage site for pesticide containers, Lagoa da Confusao, 2009.

Those responsible for the projects said that the use of pesticides didn't harm the fauna or the people present. In order to justify their response, they said*: "It can, if you leave containers, but today people are very aware, because of IBAMA, which comes here directly." "No, because it's legal and there are inspections all the time"; "It can harm over time, but it doesn't have to*

"No. When you play there's no one inside, only in the demarcated area, near the headquarters and the river you can't play. With the plane you have to stay 500 meters away and try to play at less hot and windless times because you'll go away and you won't get anything out of it"; "After they started working there, there was an increase in the number of animals in the projects of various species, which can be seen during the day and at night too".

From the interviewees' statements, it was possible to see that, although they say it doesn't harm them, they are aware that it can affect the environment in some way. On a few occasions, local hunters were seen protecting the harvest and taking the opportunity to supplement their diet with the animals they had killed.

On one of the properties visited, a dead amphibian was found stuck on the edge of a well from which water was taken to prepare pesticides (Figure 13).

Figura 13: Well from which water is drawn to prepare pesticides, Lagoa da Confusao, 2009. Source: photo by the

author, 2009.

The irrigation canals (Figure 14) have warm, slow-moving waters, which provide a suitable habitat for breeding animals, which are hosts of some diseases, such as schistosomiasis. In winter, because of the low temperature, the hosts tend to eliminate the vectors, significantly increasing the possibility of contamination of the population that comes into contact with these waters; the climate of Tocantins provides ideal conditions for the host. The disease causes extreme fatigue and lassitude in contaminated individuals, which can reduce agricultural production when it was intended to increase it (TAYLOR, 1978).

Figura 14: Irrigation canal, hydro-agricultural project, Lagoa da Confusao, 2009.

With plants to feed on, the insect begins to multiply earlier in the year, which gives hunters a better chance of catching it. In short, many pest problems have been caused by the farmer himself. When you simplify an ecosystem without prior environmental impact studies and remove undesirable species, you are taking a risk where you could lose more than you gain (TAYLOR, 1978).

The environment has the capacity to organize itself more efficiently and perfectly than human understanding recognizes. A given area can support a number of animals for a certain amount of time. In the same way, a river also has its carrying capacity, meaning that if man pollutes a river or the air, he suffers a small loss, but gains free disposal of his waste. And so exploitation tends to progress until it ruins the common property, including the exploiter himself (TAYLOR, 1978).

According to OST (1995), one of humanity's greatest presumptions is that it wants to regulate nature, where the ideal would be to regulate its own relationship with nature, agricultural methods, production techniques, habitat, energy consumption and so on. In this way, ecology can demonstrate that the health of ecosystems is a priority, more than any other short-term economic benefit, because as biological creatures, human beings, like everyone else, are deeply interdependent in the web of life.

This concept is clearly identifiable along the Urubu River. Irrigation canals have been set up and draw water from the river practically all year round. However, in summer the situation becomes more critical, when the flow decreases considerably, giving the impression that any year it could dry up completely (Figures 14 and 15).

Figura 15: Urubu River, site with inoperative dam, Lagoa da Confusao - TO, 2009.

According to AMSTALDEN (1993), pesticides released into the environment travel along many paths, concentrating at one point or another, causing contamination and the death of other species, following water or air currents, passing through organisms and causing changes in species.

The fear of losing production leads to the initiative of alternative biological pest control practices, which are not widely accepted. Many people don't know about the existence of these practices and others don't use them because they don't believe in them. In the face of any threat to production, they end up resorting to using chemical products.

During the interviews conducted with farmers in Lagoa da Confusao - TO, in 2009, pesticide sales points were also visited which supplied products that were probably sold to farmers in the agricultural projects in the study region. At the time, the municipality had a point of supply for these products, which received empty containers.

Some of the pesticides marketed by retail outlets in 2009 that could be identified are described in Table 1.

Table 1: Survey of pesticides - TO, 2009 **Source:** (Agrochemical retailers, 2009).

Item	Active ingredient	Trade name	Class	Toxicological Class (stripe)
1	**2,4 - D**	**championship**	**herbicide**	**I-extremely toxic**
2	**abamectin**	**vertimec**	**insecticide**	**I-extremely toxic**
3	**acephate**	**orthene**	**insecticide**	**III-medium toxic**
4		**cefanol**		**III - immediately**
				toxic
5	**phenoxyacetic acid**	**2.4-d**	**herbicide**	**I-extremely toxic**
6	**atrazine**	**gesaprim**		**III -medium toxic**
7	**azoxystrobin**	**priori**	**fungicide**	**III -medium toxic**
8		**prism**	**insecticide**	**III-medium toxic**
9	**beta-cyfluthrin**	**turbo**	**insecticide**	**II-highly toxic**
10	**bifenthrin**	**starion**	**insecticide**	**II - highly toxic**
11	**bispyridinyloxybenzoic acid**	**nominee**	**herbicide**	**III-medium toxic**
12	**carbedazin**	**derosal**	**fungicide**	**III-medium toxic**
13	**carboxin**	**vitavax**	**fungicide**	**III-medium toxic**
14	**ethyl carfentrazone**	**aurora**	**herbicide**	**IV - not very toxic**
15	**chlorpyrifos**	**astro**	**insecticide**	**II-highly toxic**

16	**sialofop-butyl**	**goal**	**herbicide**	**III -medium toxic**
17	**cypermethrin + thiamethoxam**	**full engeo**	**insecticide**	**II-highly toxic**
18	**clethodim**	**select**	**herbicide**	**II-highly toxic**
19	**clomazone**	**gamit**	**herbicide**	**III-medium toxic**
20	**chlorfluazuron**	**atabron**	**insecticide**	**IV-slightly toxic**
21	**deltramitrin**	**decisive**	**insecticide**	**III-medium toxic**
22	**diquat dibromide**	**reglone**	**herbicide**	**II-highly toxic**
23	**paraquat dichloride**	**gramoxone**	**herbicide**	**I-extremely toxic**
24	**difenoconazole**	**score**	**fungicide**	**I-extremely toxic**
25	**endosulfam**	**captus**	**herbicide**	**I - extremely toxic**
26		**thiodan**	**insecticide**	**I - extremely toxic**
27	**epoxiconazole + pyraclostrobin**	**operates**	**fungicide**	**III-medium toxic**
28	**spiromesifen**	**oberon**	**insecticide**	**III-medium toxic**
29	**soybean oil methyl ester**	**aureus**	**min. oil**	**IV- not very toxic**
30	**etefon**	**ethrel**	**regulator**	**I-extremely toxic**
31	**ethiprole**	**curbix**	**insecticide**	**III-medium toxic**
32	**ethoxysulfurom**	**gladium**	**herbicide**	**III-medium toxic**
33	**fenbutatin oxide**	**panther**	**acaricide**	**I-extremely toxic**
34	**phenoxaprope-p-etflico**	**podium**	**herbicide**	**II-highly toxic**
35		**starice**		**II-highly toxic**
36	**fipronil**	**standak**	**insecticide**	**II-highly toxic**
37	**Aumioxazin**	**Aumizim**	**herbicide**	**IV- not very toxic**
38	**Aunquiconazole**	**attentive**	**fungicide**	**III-medium toxic**
39		**connect**	**insecticide**	**III-medium toxic**
40	**aluminium phosphide**	**gastoxin**	**insecticide**	**I-extremely toxic**
41	**glyphosate**	**glyphotal**	**herbicide**	**IV - not very toxic**
42		**gliz**		
43		**roundup wg**		
44		**roundup transorb**		
45		**roundup ultra**		
46	**imazetapir**	**dynamaz**	**herbicide**	**I-extremely toxic**
47	**imidacloprid**	**confidor**	**insecticide**	**III-medium toxic**
48		**cropstar**		
49	**imidazolinone**	**pivot**	**herbicide**	**III-medium toxic**
50	**Iactofem**	**snake**	**herbicide**	**III-medium toxic**
51	**metamidophos**	**tamaron**	**insecticide**	**I-extremely toxic**
52	**oxime methylcarbamate**	**larvin**	**insecticide**	**II-highly toxic**
53		**tarzan**	**herbicide**	
54	**methyl metisulforom**	**ally**	**herbicide**	**III-medium toxic**
55	**methomyl**	**mentox**	**insecticide**	**I-extremely toxic**

56	mineral oil	iharol	min. oil	IV- not very toxic
57		joint oil	herbicide	
58	oxadiazone	ronstar	herbicide	III - moderately toxic
59	oxifluorfen	clincher	herbicide	III - moderately toxic
60	parathion-methyl	methomex	insecticide	I - extremely toxic
61	permethrin	pounce	insecticide	III-medium toxic
62		talcord		
63	pyraclostrobin	comet	fungicide	II-highly toxic
64	pyrazosulfurom-ethyl	sirius	herbicide	III-medium toxic
5	propiconazole + trifloxystrobin	stratego	fungicide	II- highly toxic
66	quizalofape-p-etilico	targa	herbicide	III-medium toxic
67	sulfonylurea	chlorimuron	herbicide	III-medium toxic
68	tebuconazole	folicur	fungicide	IV-slightly toxic
69		constant		
70	thiophanate-methyl	cercobin	fungicide	IV- not very toxic
71	triazine	atrazine	herbicide	III-medium toxic
72	Tricyclazole	Bim	Fungicide	II-highly toxic
73	Trifloxystrobin + tribuconazole	Native	Fungicide	II-highly toxic
74	Triflumurom	Certero	Insecticide	IV- not very toxic

The commercial success of pesticides has brought with it economic, social and environmental imbalances. The increase in the area planted with soy in Brazil has resulted in the incorporation of virgin land into production, as well as the replacement of other crops with soy. In addition, inappropriate intensive cultivation practices have caused serious environmental degradation, such as erosion and the loss of fertile soils, the silting up and pollution of important watercourses, the disappearance of springs and the loss of biodiversity (MIRANDA, 2007).

Of the classes of pesticides mentioned, insecticides are the ones that offer the greatest potential for acute health problems in the majority of agricultural regions in Brazil and worldwide, with herbicides being cited as the most widely used pesticides.

In Brazil, legislation prohibits the use of mutagenic (M) and/or carcinogenic (C) pesticides, but according to the literature, many are still used, such as the following active ingredients, among those marketed where the survey was carried out, as cited by BEDOR (2008):

- Acephate (M) (HOUR, *et al.*, 1998);
- Bifenthrin (M) (BUTLER; COHEN; SQUIRE, (1997); GOTO *et al.*, (2004));
- Cypermethrin (M) CHAUHAN *et al.*, (1997); GIRI *et al.*, (2003); L'VOVA, (1984);
- Deltamitrin (M) AGARWAL *et al.*, (1994); CHAUHAN *et al.*, (1997);
- Endosulfam (M) L'VOVA, (1984);
- Glyphosate (C and M) EL - MOFTY; SAKR, (1988); MONROY *et al.*, (2005);
- Methamidophos (M) AMER; SAVED, (1987);
- Methomyl (M) HEMAVATHY; KRISNAMURTHY (1987);

- Parathion-methyl (M) WAGNER; MARENGO; PLEWA, (2003).

The survey carried out in the farm store found 75 products sold in the region, of which 31 are herbicides, 26 insecticides and 12 fungicides and five others.

Of these products, when asked about the pesticides they used, the interviewees mentioned a few, the names of which coincided with only nine of the products they used on their crops.

Despite the fact that all those interviewed said they had technical support for the use of the products, including at the time of purchase and throughout the application, as well as whenever necessary during the planting period, it could be seen that the use of agrochemicals was rampant in the region, since products are sold that are considered mutagenic and carcinogenic by law, as highlighted above. At times, it became clear that the sellers were not properly trained to help with purchases, showing insecurity and fear when passing on information, even when an agricultural technician was present.

Humans, other animals, plants and the air/soil/water can all be affected by the effects of pesticide use, either by the movement of some of the pesticides through the wind, water and contaminated food or by the constant deviations from spraying that occur in each crop cycle. In addition, the intensive use of pesticides can lead to the illness and extinction of animal and plant species, as well as an increase in resistant pest populations (PIGNATI *et al.*, 2007).

For example, one of the pesticides used in the study region, glyphosate, effectively controls a wide variety of grasses and dicotyledonous plants (SILVA *et al.*, 2005), where its degradation is mainly carried out by microorganisms, which sometimes prevents it from reaching animals.

According to BRAGUINI (2005), aminomethylphosphonic acid (AMPA) is the first and main product resulting from the degradation of the herbicide glyphosate released into the soil, which is highly soluble in water and can enter aquatic environments through surface runoff (LIOI *et al.*, 1998), and the occurrence of glyphosate has been detected in irrigation and drainage water from rice fields in Rio Grande do Sul.

According to the classification of the US Environmental Protection Agency (EPA), glyphosate is included in group D, which means that "the agent is probably not carcinogenic to humans", based on evidence of oncogenicity in animals (BRAGUINI, 2005). The acute toxicity of glyphosate is considered low. According to the World Health Organization (WHO, 2008), the oral lethal dose 50 (LD50) of pure glyphosate in rats is 4,230 mg kg-1, (SILVA *et al.*, 2005).

In the degradation of glyphosate, the metabolite AMPA is more harmful than glyphosate itself and was found in carp 90 days after the application of the herbicide. A study at UNICAMP showed that 61% of pesticide poisonings in Brazil between 1996 and 2000 were due to manipulations with glyphosate (BOLOGNESI, 2003).

Although glyphosate is considered a low-toxicity herbicide, several cases of poisoning have been reported. Among these are some intentional cases, where the effects observed were bleeding and erosion of the gastrointestinal tract, as well as alterations in other organs such as the lung, liver, cardiovascular system, renal system and central nervous system. The estimated amount of Roundup ingested by the people who did not survive was 184 ± 70 mL. Most deaths occurred within a few hours of ingesting the herbicide (SILVA *et al.*, 2005; TALBOT *et al.*, 1991; MCCONNEL and HOSSNER, 1989).

In the region of Lagoa da Confusao - TO it was identified that some of the agrochemicals that are used and resold for use in agricultural projects have their use restricted or prohibited, due to their carcinogenic and mutagenic risk.

In humans, according to WILLIAMS *et al.* (2000), aspects related to oral and dermal absorption, bioaccumulation in tissues, genotoxicity, DNA damage and concluded that within the established standards, the herbicide did not pose a risk to human health.

However, LIOI *et al.* (1998) found cytotoxic activity and stimulation of the activity of the enzyme glucose-6-phosphate dehydrogenase in a culture of bovine lymphocytes, suggesting that glyphosate induces oxidative stress or a mutagenic effect in this species. The results obtained by PELUSO *et al.* (1998) and HIETANEN *et al.* (1983), in which the authors verified the induction of DNA adduct formation in the kidney and liver of mice and an increase in hepatic catalase activity in rats treated with commercial glyphosate, also reinforce the hypothesis of oxidative stress induction. A decrease in hepatic activity was also observed in rats treated with glyphosate (Roundup®).

The awareness that each interviewee is responsible for the consequences of pesticide use is fundamental to achieving more pragmatic solutions. The central issue is to understand that all living beings have the right to life and that they are responsible for it. It was interesting to see how everyone expressed a sense of respect for the environment, whether consciously or unconsciously, on an individual level, even if only because they know that the success of their work depends on it. However, even though they know how important it is, they adopt practices and/or products that can cause irreparable damage to the ecosystem. Most of the time, they even use them because they don't have the proper knowledge of sustainable alternatives, but another preponderant factor is the search for greater profitability, regardless of the environmental impact.

The information contained in this research will serve as a basis for future studies that can be developed in the state and especially for the region studied, as well as helping in environmental education work that also focuses on production practices that offer less degradability of natural resources, which need to be developed with agricultural producers in the region.

4 - CONCLUSION

The pesticides used in the irrigation projects visited, located in the municipality of Lagoa da Confusao- TO, are permitted by Brazilian legislation, except for nine active ingredients: Acephate, Bifenthrin, Cypermethrin, Deltamitrin, Endosulfam, Glyphosate, Metamidophos, Methomyl and Parathion-methyl.

It would be of fundamental importance to carry out more specific studies in various environmental compartments, especially in the water of the Urubu River, since it is one of the most susceptible environments to contamination, as well as facilitating the transportation of contaminants to fauna, such as *L. latrans.* One of the studies indicated for this case is the chromatography of the water and blood of these animals, which makes it possible to identify which substances are causing the alterations found in *L. latrans.*

BIBLIOGRAPHICAL REFERENCES

AGARWAL, B. **The Gender and environment debate: lessons from India. Feminist Studies**, 18 (1), 1992.

ALIER, J. M. **O ecologismo dos pobres: conflitos ambientais e linguagens de valoraçâo** (translator Mauricio Waldman). - Sao Paulo: Contexto, 2007.

ALVES FILHO, J. P. **Uso de agrotóxicos no Brasil: controle social e interesses corporativos**. Sao Paulo: Annablume; Fapesp, 2002. 188 p.

AMER, S. M., SAYED, M. A. **Cytogenetic effects of the insecticide methamidophos in mouse bone marrow and cultured mouse spleen cells.** Zeitschrift fur Naturforsch, Tubingen, v. 42, n. 1-2, p. 21-30, jan./feb. 1987.

AMSTALDEN, L.F.F. **Environment, Pesticides and Contamination: The many faces of one problem.** Revista da Associaçao Brasileira de Reforma Agrària - ABRA. Sao Paulo,01(23): 87-99, Jan./Apr. 1993.

ARANGO, H. G. **Biostatistics: theoretical and computational.** 2 ed. Rio de Janeiro. Guanabara Koogan, 2005.

BEDOR, C. N. G. **Estudo do potencial carcinogênico dos agrotóxicos empregados na futicultura e sua implicaçâo para a vigilância da saúde.** Thesis (Doctorate in Public Health) - Aggeu Magalhaes Research Center, Recife, 2008.

BERNARDO, S. **Impacto Ambiental da irrigaçâo no Brasil.** Rev. Engenharia na Agricultura - Série Irrigaçao e Drenagem. Vol. 1, no 1. Viçosa, MG; Department of Agricultural Engineering, 1992. 7p.

BOLOGNESI, C. **Genotoxicity of pesticides: a review of human biomonitoring studies.** Mutation. Research, Amsterdam, v. 543, n. 251-27. Jun. 2003.

BRAGUINI, W. L. **Effects of Deltamethrin and Glyphosate on Parameters of Mitochondrial Energetic Metabolism, on Artificial and Natural Membranes in "in vivo" Experiments.** 2005, 191 f. Dissertation (Doctorate in Biochemical Sciences) - Federal University of Paraná, Curitiba, 2005.

BUTLER, W. H.; COHEN, S. H.; SQUIRE, R.A. **Mesenchymal tumors of the mouse urinary bladder with vascular and smooth muscle differentiation.** Toxicologie pathologie, Paris, v.25, n. 3, p. 268-74, mai-jun. 1997.

CAMPOS, L. F. L. **Métodos e técnicas de pesquisa em Psicologia.** Campinas, SP: Alinea, 2004.

CHAUHAN, L. K. S., AGARWAL, D. K., SUNDARARAMAN, V. **In vivo induction of sister chromstidbexchange in mouse bone marrow following oral exposure to commercial formulations of alpha-cyano pyretroids.** Toxicology. Letters, Amisterdam, v. 93, p. 153-157.1997.

COSTA, M.A.G. **Environmental pollution:** a **legacy for future generations.** Santa Maria, RS: Orium, 2004. 254 p.

COUTO, J. L. **Riscos de Acidentes na Zona Rural.** Rio de Janeiro, UFRRJ (Federal Rural University of Rio de Janeiro) 1998. 48 p.

DORADO, A. J. **An initial assessment of agricultural colonization in Rondônia.** Campinas: EMBRAPA-NMA, Aug. 1999. 28p.

EL-MOFTY, M. M. et al. **Carcinogenic effect of gibberellin A3 in Swiss albino mice.** Nutrition and cancer, Hillsdale, v. 21, n. 2. p. 183-90. 1994

FERRARA, Lucrecia D'Alessio. The **Peripheral Look: Information, Language, Environmental Perception.** Sao Paulo: Edusp, 1993.

GRISOLIA, C. K. **Pesticides - mutations, reproduction and cancer.** Brasilia. Editora Universidade de Brasilia, 2005. 392p.

GOTO, T. *et al.* **Evolution of the Color-radius and Morphology-radius Relations in SDSS Galaxy Clusters.** MNRAS, 348, 515. 2004.

HEMAVATHY, K. C., KRISHNAMURTHY, N. B. **Evaluation of lannate 20, a carbamatepesticide in the germ cells of male mice.** Environmental Research, New York, v. 42, n. 2, p.362-365, Apr. 1987

HIETANEIN, E; LINNAINMAA, K.; VAINIO, H. **Effects of Phenoxyherbicides and Glyphosate on the Hepatic and Intestinal Biotransformation Activities in the Rat.** Acta Pharmacol. Toxicol., Copenhagen, v. 53, n. 2 p. 103 - 112, Jan. 1983.

HOUR, T. C., CHEN, L., LINDA L, LIN, J. K. **Comparative investigation on the mutagenicities of organophosphate, phythalimide, pyrethroid and carbamate insecticides by the Ames and lactam tests.** Mutation research, Amsterdam, v. 13, n. 2, p. 157-166. 1998.

LIOI, M. B. *et al.* **Cytogenetic Damage and Induction of Pro-oxidant State in Human Lynphocyte Exposed in vitro to Gliphosate, Vinclozolin, Atrazine and DPX-E9636.** Environ Mol Mutagen, New York, v. 32, n. 1 p. 39-46, Jan. 1998.

L'VOVA, T. S. **Mutagenic action of 5 prospective pesticides on mouse bone marrow, in a culture of human peripheral blood lymphocytes and on saccharomycete yeasts.** TSitologiia i genetika, [S.l.], v. 18, n. 6, p. 455-457, jà./fev. 1984

MARSH, G. P. (1847) **Where Wisdom Begins.** Article in UDALL, S. L. A crise silenciosa: A tragèdia do desmatamento e da erosao. Rio de Janeiro. 1963.

MARTINS, A. K. E. **Ipucas da Planicie do Araguaia, state of Tocantins: physical environment of occurrence, soils and land use.** 2004. 168p. Thesis (Doctorate in Forest Science) - Federal University of Viçosa, Viçosa, MG.

McCONNEL, J. S.; HOSSNER, L. R. **X-ray Diffraction and Spectroscopic Studies of Absorbed Glyphosate.** Journal Agic. Fod Chem., Washington, v. 37, p.555-560, July, 1989.

MEDEIROS, D. H de; ROCHA, J. A. da; LIMA, M.; SANTOS, R. de J. C.; ALMEIDA, A. C. G. de.; LEANDRO, Z. F. **Relationship between man and nature from the perspective of interdisciplinarity**. Campo Mourao, PR: Fecilcam, 2008.

MEIHY, J. C. S. B. **Manual de Història Oral.** Sao Paulo: Ediçoes Loyola, 2002.

MINAYO, M. C. de S. **O desafio do conhecimento**. 7. ed. Sao Paulo: Hucitec, 2000.

MIRANDA, A.C.; MOREIRA, J.C.; CARVALHO, R.; et. al. **Neoliberalism, pesticide use and the crisis of food sovereignty in Brazil.** Rio de Janeiro: Rev. Ciência & Saùde Coletiva, Rio de Janeiro, n. 12, n.1, p.7-14, jan./mar. 2007.

MONROY, C. M. et al. **Cytotoxicity and genotoxicity in human cells exposed in vitro to glyphosate.** Biomédica. Bogotà, v. 25, p. 335-345. 2005.

MONTENEGRO, R. M.; FRANCO, M. **Pathologia: processos gerais.** 4. Ed. Sao Paulo: Editora, Atheneu, 1999. p.320. 2006.

OST, F. **A natureza a margem da lei: a ecologia a prova do direito.** Lisbon: Instituto Piaget, 1995.

PASCHOAL, A. D. **Receituàrio agronòmico: fatores determinantes e limitantes.** Sao Paulo: Agroediçoes, 1982. P. 165 - 6.

PELUSO, M. *et al.* **32P-postlabeling Detection of DNA Adducts in Mice Treated with the Herbicide Roundoup.** Environ. Mol. Mutagen, New York, v. 31, n. 1 p. 55-59, Jan. 1998.

PIGNATI, W. A.; MACHADO, J. M. H.; CABRAL, J. F. **Accidente rural ampliada: o caso das "chuvas" de agrotóxicos sobre a cidade de Lucas do Rio Verde - MT.** Rev. Ciência & Saùde Coletiva, Rio de Janeiro, v. 12, n. 1, p.105-114, jan./mar. 2007.

RODRIGUES, E.L.; FANTA, E. **Liver histopathology of the fish Brachydario rerio after acute exposure to**

sublethal levels of the organophosphate Dimetoato 500. Revista Brasileira de Zoologia. Curitiba, v. 15, p. 441-450, 1998.

SAVIANI, D. **Pedagogia Histórico-critica: primeiras aproximçôes.** 5 ed. Campinas, SP: Autores Associados, 1995.

SILVA, A. M. *et al.* **Survey of pesticide use in the Bebedouro irrigation project**. 8th Northeastern Congress of Ecology. 1999. Available at: http://www.sne.org.br/congressos.html Accessed on: April 3, 2005.

SPADOTTO, C. A. **Interdisciplinary Approach to the Environmental Assessment of Pesticides**. Petrópolis: Vozes, 2006, p. 56 - 59.

STRACHMAN, M.; TAMBELINI, M. **A percepção dos Pequenos Agricultores da regiâo de Araraquara.** In: II Encontro Anual da ANPASS, Brasilia-DF, 2004. Available at http://www.anppas.org.br/encontro anual/encontro2/GT/GT05/marina strachman.pdf Accessed on 10/03/2008.

TALBOT, A. R. *et al.* **Acute Poisoning with a Glyphosate-surfactant Herbicide ("Roundup"): a Review of 93 cases.** Human Exp. Toxicology, Basingstoke, v.10, n. 1, p. 1 - 8. Jan. 1991.

TAYLOR, G. R. **A ameaça ecològica**. ed. Brasileira rev. e prefacada por Màrio G. Ferri; tradución de Rodrigo Machado. Sao Paulo: Verbo: Ed. da Universidade de Sao Paulo, 1978.

TOCANTINS. **Formoso River Basin Plan - PDH Rio Formoso, in the State of Tocantins**. Palmas: SRHMA, 2007.

UDALL, S. L. **The quiet crisis: The tragedy of deforestation and erosion.** Original North American title: The quiet crisis. Rio de Janeiro. 1963.

UNESCO. **Expert Panel on Project 13: Perception of Environmental Quality, Final Report/Programme on Man and the Biosphere (MAB)**. Paris: UNESCO, 1973.

WAGNER, E. D.; MARENGO, M. S.; PLEWA, M. J. **Modulation of the mutagenicity of heterocyclic amines by organophosphate insecticides and their metabolites.** Mutation Research, Amsterdam, v. 536, p. 103 - 115. 2003.

WILLIAMS, M. G.; KROES, R. MUNRO, I. C. **Safety evaluation and risk assessment of the herbicide roundup and its ingredient glyfosate, for humans.** Regulatory toxicology and pharmacology. Duluth, v. 31, p. 117-165. 2000.

WHYTE, A. V. T. **Guidelines for fields studies in environmental perception**. MAB Technical Notes 5, Paris: *UNESCO*, 1977.

WHO. **Deltamethrin, Environmetal health Criteria 97: International Program on Chemical Safety.** Geneva: World Health Organization, 1990. p. 2. Available at: <http://www.who.com/planet/deh97.html> Accessed on: 14 Apr. 2008.

ANNEXES

ANNEX I: Consent form issued by the Ethics Committee, CEULP - ULBRA, Palmas - TO.

UFT

FEDERAL UNIVERSITY OF TOCANTINS
MASTER'S DEGREE IN ENVIRONMENTAL SCIENCES
PALMAS CAMPUS

INFORMED CONSENT FORM

Me, undersigned,

I agree to participate in the research entitled **"Evaluation of the possible contamination of water associated with the presence of a certain root along the Urubu River, Lagoa da Confusao - TO"**, which will be carried out on the **Urubu River.** The aim is to **study the possible contamination of water caused by pesticides in two areas (one characterized by the presence of agricultural plantations and the other by the presence of native vegetation), along the Urubu River, verifying the occurrence of a frog known as *Leptodactylus latrans*, Lagoa da Confusao - TO.** The researcher will keep the information strictly confidential and will ensure my anonymity when publishing the results of the research, as well as giving me permission to withdraw at any time, without any damage to the quality of the service provided to me.

The research will be carried out by **Raquel Aparecida Mendes Lima, a** Master's student in Environmental Science, under the supervision of **Dr.ª Adriana Malvasio, a professor at the Federal University of Tocantins,** and will not involve any risk to the research subject.

I have been informed that I can contact the researcher if I wish to ask any questions about the research, by telephone **(63) 3232-8175,** address: **Av NS-15, ALC NO 14, 109 Norte, Palmas-TO, Caixa Postal 114, CEP 77001-090** and that, if I am interested, I can receive the results of the research when they are published. This research corresponds to and meets the ethical and scientific requirements indicated in CNS Resolution No. 196/96, which contains the guidelines and regulatory standards for research involving human beings.

This consent form will be kept by the researcher and under no circumstances will it be made known to anyone else.

Palmas-TO, //2009 _____ .

Participant's signature ____

Raquel Aparecida Mendes Lima — Dr.ª Adriana Malvasio

Academic-Researcher — Teacher Advisor

ANNEX II: Script for interviews with farmers in Lagoa da Confusao - TO, 2009.

Interview n°

1 - Location (municipality and geographical coordinates of the property):

2 - Property Name: ____

3 - Type of property: ( ) Private ( ) Leased

4 - Age:5. Sex:6. Educational level:

7 - Position/activity involved:

( ) Preparation ( ) Application ( ) Cultivation ( ) Transportation

8 - Property area:

9 - Legal Reserve Area:

10 - There is a degraded area of permanent preservation:

11 - Type of exploitation:

( ) Agriculture ( ) Livestock ( ) Conservation ___________________

Item	Planted area	Product	Crop cycle	Productivity
1				
2				

12 - Type of planting: ( ) Direct ( ) Conventional

13 - Use of pesticides:

Planted product:					
Trade name	Form of application	Applied dose	Time application	No. of applications per harvest	Destinations packaging

14 - Use of personal safety equipment - PPE:

15 - Reads the label of the chemical products used:

16 - You receive specialized technical assistance:

17 - Last contact period for specialized technical assistance:

18 - Did you notice any symptoms after the application process? What were they?

19 -Where does the water used for irrigation come from?

20 - Where does the water used for human consumption come from?

21 - Do you know where the water that leaves the irrigation project returns to?

22 - How many times a day are the irrigation pumps turned on?

23 - How many hours do the pumps stay on?

24 - How much water is drawn from the river, per time it is on or per hour?

25 - What time of year do you irrigate? Which months?

26 - When does the use of chemical products start?

27 - Is there an interval between applications? How long?

28 - Do you use the same chemical product more than once during the year? Which ones and how often?

29 - What are the advantages of using chemicals in the project?

30 - Are there any disadvantages or harms to using chemicals in the project? What are they?

31 - Would it be possible to produce without using the chemicals in the project? Why?

32 - Do you know of any alternative practices to reduce or eliminate pests or animals that harm the project?

33 - Do you have any other source of income apart from the project?

34 - Do you think the products used could harm anyone or any animal on the project site?

35 - Do you think that the products used could harm any animals or people who are not working on the project, but who are nearby or even on neighboring farms?

36 - How many people work on the project per season or per year?

CHAPTER III HISTOLOGICAL ANALYSIS OF THE TEGUMENT, LIVER AND REPRODUCTIVE APPAREL OF LEPTODACTYLUS LATRANS, TOCANTINS

SUMMARY

The agricultural model adopted in Brazil is based on the use of agrochemicals, which in a way is disorganized and excessive and has caused various impacts on the environment. Depending on their chemical nature and concentration, pesticides released into the environment can cause various types of damage to the biota exposed to them. Several studies have shown in different organisms, including amphibians, that some pesticides can influence the survival, fertility and genetic composition of populations. The municipality of Lagoa da Confusao is known for its large agricultural projects, and the Urubu River is one of the most degraded rivers in the region, due to the presence of large and numerous agricultural projects on its banks. Taking into account the impact caused by irrigation projects, agricultural activities and the use of pesticides, especially in that region, the objective was to analyze the living tissues (liver, epithelium and reproductive system) in order to verify the existence of alterations in *L. latrans* caught along the Urubu River, Lagoa da Confusao and Cristalândia - TO. For this purpose, histological analysis of the aforementioned tissues was used, which mainly showed significant alterations in the liver of the animals caught at the three study sites. In the liver from the three study sites, there were alterations in its composition, such as: the presence of granulations forming densely stained and poorly dispersed extracellular corpuscles, as well as more condensed corpuscles of varying sizes; compartmentalization of the hepatic lobules; intense vacuolization of the hepatocytes; relative and intense hyperemia with loss of the vascular framework and leukocyte infiltration. In the skin, the histopathological alterations found were: abundant melanin near the basal lamina; absence of a keratinized layer; basal layer with large or bulky euchromatic nuclei. Both the female and male reproductive systems showed no alterations. The existence of these alterations suggests the presence of chemical products in these areas which may cause the histological stress identified.

Keywords: tissues, histopathology, amphibians, irrigation.

1 - INTRODUCTION

Environmental pollution is increasing frequently, especially in aquatic ecosystems. Various negligent actions sometimes lead to an increase in environmental impacts. This pollution of aquatic ecosystems can lead to a loss of biodiversity, resulting in the reduction or disappearance of populations. In view of this, environments have been monitored, which is fundamental in the search for knowledge of the local reality, as well as being one of the greatest human concerns, since direct and indirect consumption of contaminated water can cause serious damage to organisms.

Amphibians were the first vertebrates to colonize the earth and their life cycle is closely linked to water. The name amphibian refers precisely to their double life (amph = both and bio = life), indicating the transition from the aquatic to the terrestrial environment. Adults with lungs live on land, but are always dependent on water to survive and perpetuate the species (AURICCHIO & SALOMÂO, 2002).

Today's amphibians belong to the subclass Lissamphibia (smooth-skinned amphibians), made up of three orders: Urodela or Caudata (Salamandra and tritoes), amphibians with two to four limbs and a long tail; Apoda or Gymnophiona (blind snakes or cecilias), amphibians without limbs, with a vermiform shape and a large number of body rings, and Anura (frogs, toads and tree frogs), amphibians without tails and with longer hind legs, usually modified for jumping locomotion (POUGH, *et al.*, 2003).

In the larval stage, they breathe through gills; as adults, they have pulmonary and cutaneous respiration. Their lungs are inefficient organs, as their large chambers, the simple alveoli, have a relatively small internal surface area for gas exchange, and air pumping into them is precarious. The skin plays an important respiratory role, guaranteeing the greatest absorption of oxygen (SILVA JR. & SASSON, 2005).

Some causes of the disappearance of species may be related to habitat alterations caused by deforestation, fires, acid rain, increased ultraviolet radiation, the presence of certain animals such as cattle, the use of acidic or toxic chemicals, among others (POUGH, 2003).

Amphibians have recently been the object of study in numerous situations, as many species are able to indicate changes in the quality of the environment. This is largely due to the fact that they live in diverse environments, both water and soil, making them indicators of environmental quality. Another important criterion to consider is the sensitivity of ras, for example, as they are vulnerable to environmental disturbances at various stages of their life cycle and can be exposed to pesticides through the food chain. While wrasse eggs have no protection, as their coats are absent, as adults they have great absorption power through their thin, permeable skin, which can immediately absorb pollutants from water, air and soil (MILLER, 1931).

Because various environmental factors affect the size of organism populations, plus the specific characteristics of amphibians, such as skin permeability and a life cycle dependent on both aquatic and terrestrial environments, these terrestrial vertebrates are more vulnerable to environmental variations. Consequently, the reduction in the amphibian population in a given region is considered a bioindicator of environmental quality (RABB, 1990; STEBBINS & COHEN, 1995; FEIO et al., 1998).

In environments where pollutants are present, the liver can develop histological changes in hepatocytes, which can be related to the effects of contaminants (FANTA et al., 2003), making it an organ of fundamental importance for investigating the action of chemical pollutants on aquatic organisms. This organ has the power to accumulate substances, leaving liver cells exposed to chemical agents present in the environment or in other organs (HEATH, 1987).

The state of Tocantins, more specifically the region of Lagoa da Confusao, known for the large number of agricultural projects that drive the region's economy, also has a rich biodiversity, with fauna populations being one of the most exuberant, but little studied, especially when we refer to histological aspects.

The digestive system of aquatic anurans uses suction to swallow food in the water, but most semi-aquatic and terrestrial species have a long tongue, attached to the anterior region of the mandible and which can be projected outwards to catch insects, and both have wide mouths without teeth. In the terminal region of the body is the cloaca, where the genital, urinary and digestive canals end (SILVA & SASSON, 2005; POUGH et. al., 2003).

Monitoring the effects of pollutants on amphibians is of great importance when assessing the stress caused by pollution. This study sought to identify changes in *L. latrans* and correlate them with environments where pesticides were used, in turn comparing them to another environment that appeared to be free of pesticides. A ra is an anuran that, after its evolution in water, shows a predilection for aquatic life and lives on the edge of ponds, waterholes and other collections of water. It has moist skin, a slimmer body and longer legs than frogs, which are well adapted to jumping (SANTOS, 1994).

Histological study can help to understand various biological and ecological processes, such as: tissue analysis, by analyzing the microscopic composition and the respective function of the biological material, it also includes the structure of cells and organs, trying to explain the interrelationships between cells, tissues and the structure and composition of organs, as well as representing the link between biochemistry, physiology and genetics, on the one hand, and pathology and the clinic, on the other (JUNQUEIRA and CARNEIRO, 2008; GENESER, 1999).

To this end, histological studies were carried out on the tissues, liver, skin and reproductive system of *L. latrans, in* the Urubu River, Lagoa da Confusao and Cristalândia, in an attempt to identify histopathological alterations associated with hygroagricultural project sites, *since* amphibians are considered good bioindicators of contamination due to the fact that they live in terrestrial and aquatic environments during a certain phase of their lives and because they are thermoregulators.

2 - MATERIAL AND METHODS

2.1 - Request for authorization from IBAMA (License)

All collections were carried out following the protocols authorized by IBAMA under process number 16736, as shown in Annex I.

2.2 - Choice of species

Many authors see amphibians as the result of specific responses to environmental characteristics and reflect the correspondence of independent life histories in time and space (GASCON, 1991). It would therefore be possible to investigate pesticide contamination in the Urubu River indirectly over a short period of time, i.e. six random months of data collection.

Some physiological characteristics (permeable skin) and ecological characteristics (biphasic life cycle) make amphibians strongly dependent on water. They are highly sensitive to changes in the physical and chemical parameters of water, and several species are sensitive to changes in the structure of vegetation in the vicinity of water bodies (JIM, 1980; VAN DAM and BUSKENS, 1993; BURKETT and THOMPSON, 1994; WATSON *et al.,* 1995). This information is essential to help develop the study of the desired environment.

The fact that they are abundant and functionally important in many terrestrial and aquatic habitats in tropical, subtropical and temperate regions, they are significant components of the Earth's biota, which allows them to have a wide distribution and potentially serve as key species to assess changes in the environment, as well as other species that are habitat specialists or have a restricted distribution, and can indicate a local disturbance (HEYER *et al*., 1993). The dependence of amphibians on aquatic and terrestrial environments makes them an excellent source of environmental indicators, since any disturbance caused in these environments can directly affect the balance of amphibian species (DIXO, 2001).

In this sense, *L. latrans, because it is an* Anuran, meets the descriptions needed to be the indicator of contamination that this research is aimed at, as it also shows a predilection for aquatic life and lives on the edge of ponds, waterholes and other water collections, making it easy to identify its presence in the places chosen for the study, in addition to being present in agricultural projects, due to the existence of irrigation canals.

2.3 - Location of collected material - *L. Iatrans*

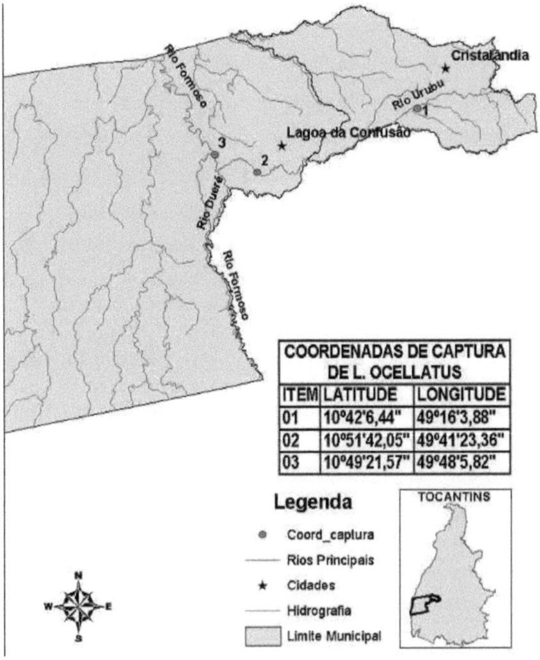

Figure 1: Map showing *L. latrans* capture sites, Cristalândia/ Lagoa da Confusão � TO, 2009. Source: BASTOS, 2010.

During the environmental reconnaissance phase, sites were visited along the banks of the Urubu River, sampling the species' areas of occurrence. The locations marked as capture points are shown in Figure 1, symbolized by the green circles, with numbers corresponding to 1, 2 and 3.

2.4 - Procedures for preparing the animals in the field

2.4.1- Materials and equipment used during catches

The capture procedures were based on AURICCHIO & SALOMÂO (2002), as described below:

The team used *Mag lite* flashlights *with* alkaline batteries to make the captures.

In addition, it was necessary to have pincers, which were used to capture larger individuals that were in places that were difficult to access, such as the banks of canals, thus preventing the specimens from escaping, and for the other animals, the capture was carried out manually.

2.4.2- Collections

Six samplings were carried out per point, i.e. 18 samplings, corresponding to 90 hours of captures, between the hours of 6:30 p.m. and 11:00 p.m., and a point could be visited on the same night. The capture time was usually around one hour per observer, and there were always three to six observers, giving a total of five hours of capture per point.

Capture work always began at dusk, when some species were already singing. The site was analyzed during the day to reconnoiter the area. The sites were always next to the Urubu River, and also had the presence of marshes or lagoons, and in the agricultural project area, mainly irrigation canals, as these are where amphibians usually live. In order to capture them, it was necessary to use torchlight and focus on the animal's eyes, and then capture it.

The animals were placed in plastic bags with zippers, so that they could breathe and not die until they arrived at the team's support point. The support points used to work the following day were located at the Lago Verde and Praia Alta farms, Lagoa da Confusao - TO.

As amphibians generally breed between November and March, the rainy season, they are easily found in: ponds, pools, inside bromeliads, shallow holes in the forest floor, and can therefore be captured with bare hands (AURICCHIO and SALOMÂO, 2002). Collections were carried out taking seasonality into account, during 2009, in the months of March, April, May, July, September and October.

2.4.3- Packaging, transportation after collection, sacrifice, fixation after collection, preparation, storage of collected animals and preservation of samples in 10%% formaldehyde.

The amphibians were placed in transparent plastic bags with zippers (Figure 2). A maximum of 5 (five) individuals of varying sizes were placed in the bags, so as not to cause death by drying out or injury, always blowing on the inside of the bag before placing the animal, to oxygenate the environment, during the time of capture. The species of ray studied (*Leptodactylus*) is very resistant, which is why, as a precaution, the bags containing the captured animals were always placed in a plastic container with a lid until the end of the capture, in order to prevent them from escaping.

Figure 2: *L. latrans*, caught and stored in plastic bags, TO, 2009. **Source:** Barreira, 2009.

Once they had been captured, the animals were taken to the support farm, removed from their bags and placed in plastic containers with a little water from the river under study, where they spent the night. The following morning, procedures were carried out to preserve the samples or mark the animals that were to be returned to their original environment.

In glass jars with a capacity of approximately 4 liters, a 10% formaldehyde solution was poured to preserve the animals until the next procedures, i.e. histology of the parts of interest (Figure 3).

Figura 3: Jar with 10% formaldehyde solution, with sacrificed *L. latrans*. TO, 2009.

Source: photo by the author, 2009.

The vials were all duly identified with external labels containing the following information: place of collection, date, species under study, type of solution contained in the vial. All the animals that were sacrificed carried a pencilled label at waist height, marking: the animal's number, date of capture, collection point (Figure 4).

Figura 4: Label used to identify animals preserved in 10% formaldehyde, TO, 2009. Source: photo by the author, 2009.

A maximum of ten individuals were placed in each jar, depending on the size and space, observing all the samples so that they didn't get caught in the formaldehyde solution, thus avoiding their possible loss. At the end of the catches, these jars were sent to the Ecology and Zoology Laboratory at the Federal University of Tocantins, where they were stored until the initial histological preparation.

The animals were killed, as authorized and recommended (AURICCHIO & SALOMÂO, 2002).

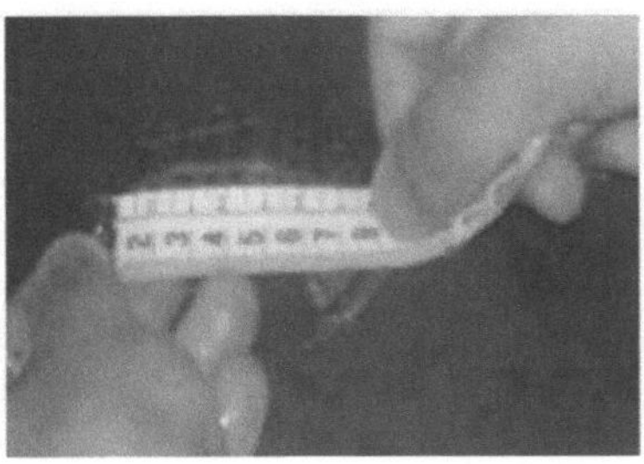

Figure 5: Measuring the animals, TO, 2009. Source: MORAES, 2009.

After the catches, the animals were separated by spot and kept in a container with a little water to avoid death, where they usually waited until the morning of the following day to continue their work.

The captured and freshly killed amphibians were washed with tap water, also from the Urubu River, to remove any kind of residue, so as not to risk damaging the preserved individuals. Then the rostro-cloacal measurements were taken, they were labeled one by one and all the individuals from a point and a catch were placed in a jar. The outside of the jar was also labeled.

The animals were chosen at random for sacrifice, analyzed and identified, duly listed in a field notebook. The animals were stored in a scientific collection at the Ecology and Zoology Laboratory of the Federal University of Tocantins, in Palmas - TO. To save space and jars, several animals were placed in a single container, in a liquid medium to prevent drying out, except for the parts that were preserved in paraffin.

2.3.4 - Marking method used

Chart 1: Marking method used for the animals captured, Tocantins, 2009.

Method	Procedure	Advantages	Disadvantages	References
Subcutaneous colored fluorescent mark	Apply a colored fluorescent mark under the skin	Easy Many individuals	Dear Temporary (1 to 2 years)	Nishikawa & Service, 1988 Emlen, 1968 and Davies & Hallyday 1979 Bailey, 2004 Johnson & Wallace 2002

The choice of marking method was based on assessing whether the method corresponded to the needs. The aim was to choose a method that was easy to work with, that could ensure marking during the study period and that was easy to identify and inexpensive. The subcutaneous colored fluorescent marking method met almost all the requirements except for the budget issue (Table 1). However, the project relied on the donation of the biopolymer marking material (Figure 6).

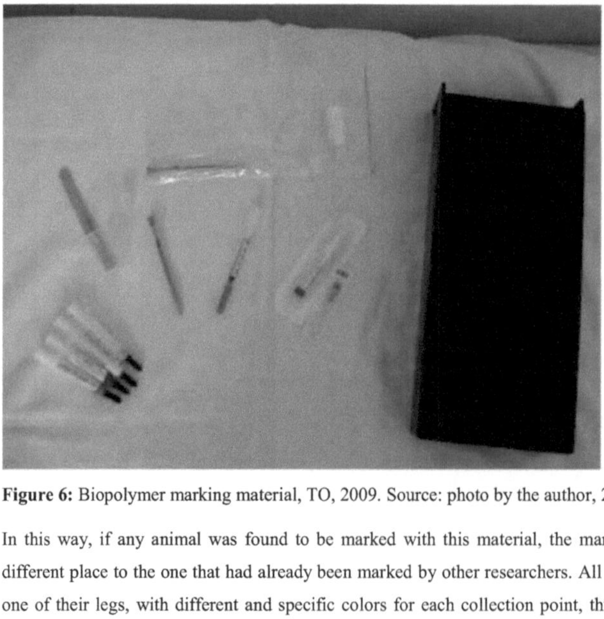

Figure 6: Biopolymer marking material, TO, 2009. Source: photo by the author, 2009.

In this way, if any animal was found to be marked with this material, the mark for this study would be made in a different place to the one that had already been marked by other researchers. All the *L. iatrans* were always marked on one of their legs, with different and specific colors for each collection point, thus distinguishing them from the other markings made, as they were on the ventral part of the individuals. For each study point a different color of marker was used, at point 1 - pink, point 2 - lime yellow and at point 3 - orange, always on the hind leg (Table 2).

Table 2: Marking colors per study point, Lagoa da Confusao - TO, 2009.

Fluorescent biopolymer brand color	Location of the mark on the individual	Capture point
Pink	Hind leg	1
Lime yellow	Hind leg	2
Orange	Hind leg	3

The individuals captured were all marked and returned to their place of origin, i.e. where they were caught, as shown in Figure 7.

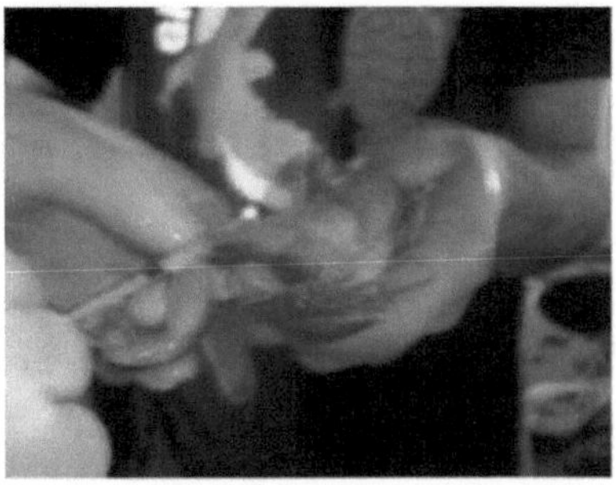

Figure 7: Marking on the hind leg of the animals to be returned to the place of capture, TO, 2009. Source: photo by the author, 2009.

2.3.5 - Amphibian sampling method

With regard to sampling methods for *L. latrans, the* decision was made not to use traps, as this is a single species study. The chance of catching only individuals of the species of interest would be low, and the work would be more exhausting. The result might have been different if the visual encounter method had been used in conjunction with the transect (Table 3) (HEIYER *et al.*, 2003).

Chart 3: Method used to mark animals captured and returned to their place of origin, 2009.

METHOD	DATA	TIME	COST	PEOPLE
Visual encounter	Relative abundance	High	low	Bass
Transect	Density and relative abundance	High	low	Medium

Figure 8: Trail reconnaissance, TO, 2009. Source: Guerreiro, 2009.

The sampling effort for each observer was a minimum of 30 minutes and a maximum of one hour, depending on how many observers were present on the day of capture (Figures 8, 9 and 10).

Figures 11 to 18 depict the reality of each point delimited for the development of this research. It should be remembered that point 1 characterizes an area with native vegetation, with the presence of a clay extraction activity, in Cristalândia - TO; point 2 and 3, with hydroagricultural projects, Lagoa da Confusao - TO.

Point 1: Native vegetation, clay extraction point, Cristalândia - TO, 2008.

Figure 11: Native vegetation - dry season, Cristalândia - TO, 2008 Source: photo by the author, 2009.

Figure 12: Native vegetation - rainy season, Cristalândia - TO, 2009 Source: photo by the author, 2009.

Point 2: Presence of agricultural plantations.

Figure 13: Agricultural planting - dry season, Lagoa da Confusao - TO, 2008.

Source: author, 2009.

Figure 14: Agricultural planting - rainy season, Lagoa da Confusao - TO, 2009

Source: photo by the author, 2009.

Point 3: Mouth of the Urubu River, Meeting with the Formoso River, Lagoa da Confusao - TO.

Figure 15: Mouth of the Rio Urubu, meeting the Rio Formoso - dry season, Lagoa da Confusao - TO, 2008. Source: photo by the author, 2009.

Figure 16: Mouth of the Urubu River, meeting the Rio formoso - rainy season, Lagoa da Confusao - TO, 2008. Source: author, 2009.

2.4 - Laboratory tissue preparation procedure

2.4.1- Preparation of tissues from individuals preserved in 10% formaldehyde

In June 2009, the Microbiology Laboratory dissected the individuals captured in the first stage (individuals preserved until June 2009, Figure 17). And the second, those preserved until November 2009.

Figura 17: Dissection of *L. latrans*, UFT, Palmas - TO, 2009. Source: photo by the author, 2009.

The first stage was carried out by two people under the supervision of the research supervisor. The tissues from each animal were preserved in a 10% formaldehyde solution and then sent to the Histopathology Laboratory at the Federal University of Goiás - UFG.

2.4.2 - Techniques for preparing and analyzing histological slides

2.4.2.1 - Slide preparation:

In order to carry out the histology, a sequence of stages was required, based on VOLNEI and SIQUEIRA (2000) (Chart 4). The slides of the animals captured in the first semester were prepared in June 2009 at the Histopathology laboratory of the Federal University of Goiás (UFG) and the second stage took place in November 2009.

The histological study was coordinated by Professor Josefa Nascimento e Lima e Lima from the Biology Course at UFT and by technician Antônio Souza da Silva from the Histopathology Department at the Federal University of Goiàs - GO.

A total of 100 animals were sacrificed and 202 histological slides were prepared from the sacrificed animals.

Chart 4: Description of the preparation of the paraffin blocks with the pieces of *L. latrans*, UFG - GO, 2009.

ACTIVITY	CONCENTRATION	TIME
Capture	10% formaldehyde	Referring to the months of capture
Fixing	10% buffered formaldehyde	24 hours
Refixing - ideal	10% buffered formaldehyde	24 hours
Washing the material in tap water to remove the formaldehyde pigments that had created crystals due to the age of the reagents.	Tap water	One night around 12 hours
Dehydration - removal of water from the part	Alcohol 70%	20 min.
	Alcohol 80%	20 min.
	Alcohol 90%	20 min.
	100% alcohol	30 min.
	100% alcohol	30 min.
	100% alcohol	30 min.
	100% alcohol	30 min.

	100% alcohol	30 min.
Clarification or diaphanization	Xylol 1	30 min.
	Xylol 2	30 min.
Impregnation	Paraffin bar, histopathological paraffin, petroleum-derived, P1004.08.AH, 56 °C 1000 g, lot 117110	
	Paraffin 1	Normally we leave it for 1 hour, but because the pieces were quite hard, we preferred to leave it for only 25 minutes.
	Paraffin 2	Normally it's left for 2-3 hours; for the same reason it was left for 25 minutes.
Inclusion	Paraffin blocks	

2.4.2.2 - Making histological slides

In the Microbiology Laboratory at the Federal University of Tocantins, Palmas - TO, the animals were dissected and placed in dorsal decubitus, opened with scissors and the tissues of interest (liver, skin and reproductive system) were removed using a scalpel and razor and stored in smaller jars in the same 10% formaldehyde solution (Figure 18).

Figura 18: jars with tissues, after dissection, Goiânia - GO, 2009. Source: photo by the author, 2009.

At the Histopathology Laboratory, UFG - GO, the tissues were washed, dehydrated in alcohol, clarified, impregnated and then embedded in paraffin blocks, as shown in Figures 19, 20, 21 and 22.

Figure 19: Preparation of paraffin blocks, Goiânia - GO, 2009. Source: photo by the author, 2009.

Figure 20: Preparation of paraffin blocks, Goiânia - GO, 2009. Source: THOMAZI, 2009.

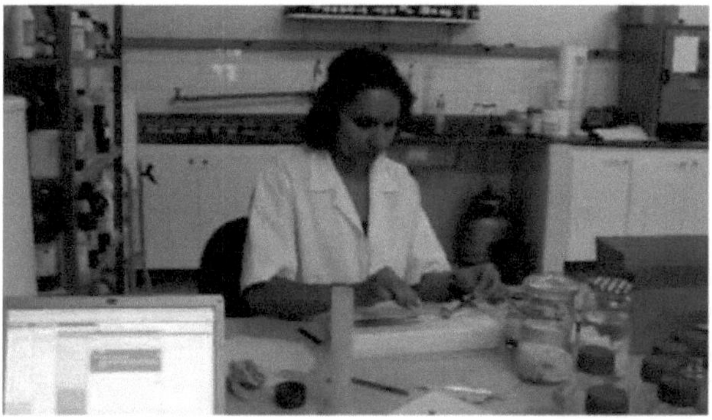

Figure 21: Tissue cut-outs for the preparation of histological specimens, Goiânia - GO, 2009. Source: Thomazi, 2009.

Figure 22: Tissue cut-outs for the preparation of histological specimens, Goiânia - GO, 2009.

Source: Thomazi, 2009.

Figure 23: Preparation of paraffin blocks, Goiânia - GO, 2009. Source: Thomazi, 2009.

In order to fix the tissues on slides, the following procedure was carried out: small fragments of the tissues were placed in paraffin blocks and subjected to very fine cuts with the aid of a cutter known as a microtome. Once cut, the paraffin was removed and placed on slides, where the cut was fixed to prevent deterioration and stained. The stain used was hematoxylin and eosin (HE).

Because parts of the skin had been removed from the legs, along with the bone structure of the *L. latrans, it* was necessary to proceed with bone wear and decalcification, according to VOLNEI and SIQUEIRA (2000).

2.4.2.3 - Material identification

The identification of the material was based on the macrostructural characteristics of the studied: liver, skin and reproductive system.

The material, i.e. the tissues, was observed under an optical microscope, which uses transparency.

Photographs were taken using a SONY model W70 digital camera, with a resolution of 7.0 megapixels, coupled to the eyepiece of an *Olympus* microscope. The lenses had magnifications of 4, 10, 40 and 100X, for which immersion oil was used.

3 - RESULTS AND DISCUSSION

3.1 - Observation of *L. latrans* along the Urubu River, Lagoa da Confusao and Cristalândia - TO, 2009

Changes in temperature affect biochemical and physiological processes (from digestion to locomotion), as well as influencing feeding and escape from predators, affecting equilibrium and being a mechanism for altering the geographical distribution of species (FILONOV, 2000). With regard to geographical distribution, it was easy to find animals at the three points analyzed, not least because it is a common species in the region.

The rostro-cloacal length of an amphibian refers to the size of the animal, portraying the stage of life it is in. The results of this biometric are shown in figure 27.

The area's climate is characterized by having two well-defined seasons: a dry one (winter) and a rainy one (summer). The driest months are from August to April and the wettest from January to October (BRASIL, 1994; TOCANTINS, 1997). When comparing the length of the animals caught during the months of April, May and July, which refer to the dry season, the animals had shorter rostro-cloacal lengths compared to the months of September, October and March (the rainy season).

The fact that the animals were larger may have something to do with their adulthood and the fact that they were breeding during the rainy season and depended on water to do so.

With regard to the comparison between the study points, it was noticed that point 1, a place with the presence of native vegetation and with the presence of clay extraction activity in the Cristalândia region, the animals were larger in relation to the other points, as shown in figure 27, highlighted in blue. Point 2 was the second place for larger animals, the area of agricultural plantations, in red, and point 3, also with the presence of agricultural plantations, was the place where smaller individuals were present.

The existence of larger or adult individuals is linked to the area of least disturbance, i.e. despite having clay extraction activities which do generate environmental impacts, the site was characterized by the presence of native vegetation, The site was characterized by the presence of native vegetation, at the headwaters of the Urubu River and in a better state of conservation than the other two points analyzed (2 and 3), with the difference that during the rainy season there was no extraction, so there was no presence of machinery on the site, making the vegetation more closed and access more difficult. At that time, the area was used for cattle breeding. Both the presence of machinery and cattle can scare away *L. iatrans*, as well as causing the death of some if they are trampled or run over by machinery. Where cattle are present, amphibian species may disappear due to changes in the environment (grass layer), with native species being replaced by exotic ones, which cause the soil to shade (SILVANO *et al.*, 2003).

Figure 28 shows the animals that were captured during the visits made to the study points in March, April, May, July, September and October 2009. According to figure 28, point 1 was the place where the fewest animals were captured during all the

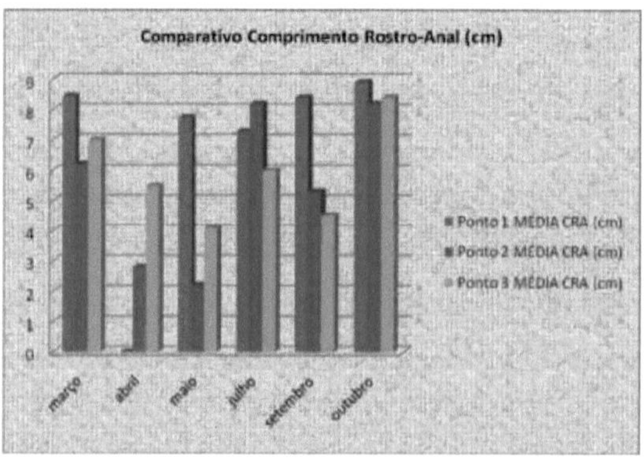

Figure 27: Ratio of the rostro-cloacal length of the animals caught, Lagoa da Confusao and Cristalândia - TO, 2009. CRA = rostro-cloacal length.

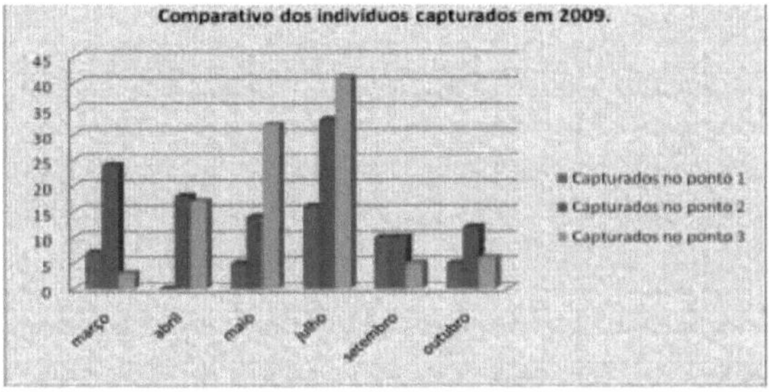

Figure 28: Number of animals caught, Lagoa da Confusao and Cristalândia - TO, 2009.

In May and July, between 30 and 41 animals were captured at point 3 per month. It was more common to find the animals in anthropized places, i.e. at points (2 and 3) where there was an agricultural project, near the irrigation canals, places where water was present.

The rainy season made capturing difficult, due to the change in climate. Any changes, such as the presence of wind or rain, caused the animals to take refuge, and they were not easily found in the places selected for the search route. There were more animals captured in areas with agricultural plantations (points 2 and 3), i.e. up to 40 animals per search night.

The effect of pesticides on each individual in the population can involve responses in terms of the abundance and stability of communities, putting biodiversity at risk (MAZON, *et al.*, 2000). In relation to the analysis carried out, it was possible to notice that in terms of the presence of the animals, it was more common to find them in areas with the presence of agricultural projects, contrary to what was perceived in relation to size or age (young and adults). The animals located in the agricultural project area were smaller and more numerous.

Seasonal variation can also influence the distribution of anuran amphibian species. Their breeding season is highly affected by the distribution of rainfall, mainly because the availability of aquatic sites for breeding is greater during the rainy season (AICHINGER, 1987). Neotropical amphibians are extremely dependent on rainfall. A single physical factor, rainfall distribution, regulates the breeding activity patterns of anurans in tropical areas that are characterized by a pronounced dry season (HEYER, 1973; ZIMMERMAN and RODRIGUES, 1990).

We must also take into account the "time" factor that this study covered, because by studying the impacts caused over the course of a year, it was possible to notice some changes in the distribution of these animals, when, in fact, it is not known that this picture could be totally different in a few years' time when the use of pesticides that are not regulated by law continues. The populations of *L. latrans* could suffer serious consequences considering the time scale.

3.2 - Histology

The general results of the histological analysis of the hepatic section, skin, female and male reproductive system of *L. latrans,* stained with HE, (Tables 5, 6, 7, 8 and respective figures: liver 29 to 34, skin: 35 to 38, female reproductive system: 39 and 40, male reproductive system: 41 to 44) showed that there were alterations in the tissue samples of some of the animals analyzed.

In order to display the information obtained in the tables, it was decided to start with the histological descriptions that were closest to the normality found, generally in a "healthy" animal towards the most altered animal, with the execution of Tables 7 and 8, on the female and male reproductive system, where only the histological description of the different stages of reproduction found is shown.

Chart 5: Histological description of the liver sections of the animals analyzed, Lagoa da Confusao and Cristalândia, 2009.

N° Figure	Làmina cuts	Histological pattern observed	Possible changes
Figure 29	24; 79; 96;	Presence of classic hepatic lobules and hepatic portal triad. Polyhedral hepatocytes organized into acini, identifiable space of Disse, filled with sinusoids. Hepatic sinusoids passing through the hepatocytes, presence of arteries in various calibres, presence of hepatic veins in various calibres, nucleated blood cells, presence of fibrous capsule formed by 3 to 4 layers of collagen fibres interspersed with spindle cells; non-vacuolized hepatocytes.	Presence of granulations forming the densely stained and sparsely dispersed extracellular corpuscles.
Figure 30	23; 29; 77; 78;	Presence of classic hepatic lobules and hepatic portal triad; polyhedral hepatocytes organized in acini, identifiable space of Disse, filled with sinusoids. Hepatic sinusoids passing through the hepatocytes, presence of arteries in various calibres, presence of hepatic veins in various calibres, nucleated blood cells, presence of a fibrous capsule formed by 3 to 4 layers of collagen fibres interspersed with spindle cells; non-vacuolized hepatocytes.	Presence of granulations forming densely stained extracellular corpuscles of larger sizes condensed.
Figure 31	22; 46; 60; 67; 86; 92; 101;	Compartmentalization of the hepatic lobules, presence of the hepatic portal triad; presence of granulations forming the densely stained extracellular corpuscles; relative cytoplasmic vacuolization of the hepatocytes.	1.Presence of granulations forming densely stained extracellular corpuscles; 2. Compartmentalization of hepatic lobules; 3. vacuolization relative cytoplasm of hepatocytes.
Figure 32	20; 94; 95;	Compartmentalization of the hepatic lobules, presence of the hepatic portal triad; large amount of granulations forming the densely stained extracellular corpuscles; intense vacuolization of the hepatocytes.	1. A large number of granulations forming densely stained extracellular corpuscles; 2. Compartmentalization of the hepatic lobules; 3. Vacuolization intense of hepatocytes.

Figure 33	26	Compartmentalization of the hepatic lobules, intense hyperemia and loss of the vascular framework; presence of granulations forming densely stained extracellular corpuscles of relatively medium to large size.	1. presence of granulations forming densely stained extracellular corpuscles of relatively medium to large size; 2. Compartmentalization of the hepatic lobules; 3 Intense hyperemia with loss of the vascular framework.
Figure 34	3; 6; 11; 49; 61; 66; 81; 82; 97;	Compartmentalization of the hepatic lobules, relative hyperemia with loss of the vascular framework and leukocyte infiltrate; presence of granulations forming densely stained extracellular corpuscles of relatively medium to large size.	1 Presence of granulations forming densely stained extracellular corpuscles of relatively medium to large size; 2. Compartmentalization of the hepatic lobules; 3 Relative hyperemia with loss of the vascular framework and leukocyte infiltrate.

Liver

According to GEORGE and CASTRO (1998), the liver is the largest gland in the body and its morphological unit is the hepatic lobe, which is grouped together to form lobes. Each hepatic lobe is represented by a parenchyma of hepatic cells (hepatocytes) and a connective-vascular stroma. The hepatocytes are arranged in radially oriented plaques made up of a single layer of cells. The hepatocytes, which are polyhedral in shape, when one of their surfaces comes into contact with that of another neighboring cell, allow a tubular space to remain between them, limited only by the cell membranes, a space which is called the biliferous canaliculi. These canaliculi extend between the plates, towards the periphery of the lobe, where they flow into a cloacal duct lined with simple columnar or cubic epithelium, the biliferous duct of the portal space.

The hepatic cells are separated from each other by sinusoidal capillaries, rich in phagocytic cells that have been given the name Kupffer's cells, belonging to the macrophage system. The sinusoid capillaries are home to terminal branches of the hepatic artery and portal vein, so that there is a mixture of arterial and venous blood flowing to the center of the lobe where the centrilobular vein is located, which collects the blood from the sinusoids. The various hepatic lobules are delimited by connective tissue. The regions rich in connective tissue and located at the corners of the lobules have been given the name of portal spaces, occupied by an arteriole (branch of the hepatic artery) a venule (branch of the portal vein), a biliferous duct and lymphatics. The main functions of the liver are: bile production, glycogen storage, lipid storage, protein synthesis (fibrinogen, prothrombin, albumin), vitamin storage (A, D, B12, K), drug metabolization, all of which are carried out at hepatocyte level (JUNQUEIRA E CARNEIRO, 2008).

In the presence of pollutants, the liver can develop histological changes in hepatocytes that can be used to monitor the effects of contaminants (FANTA *et al.*, 2003). The liver is a strategic organ when trying to understand the action of chemical pollutants on aquatic organisms. It is the first organ in the biotransformation of xenobiotics and probably also for the excretion of metals. Thus, many metals and other xenobiotics accumulate in the liver, so that liver cells are exposed to a high level of chemical agents that may be present in the environment or in other organs (HEATH, 1987).

The structure normally found in the liver, observed in *L. latrans* in the study region was: presence of classic hepatic lobules and hepatic portal triad; polyhedral hepatocytes organized into acini, identifiable space of Disse, filled with sinusoids; hepatic sinusoids passing through the hepatocytes; presence of arteries of various calibres; presence of hepatic veins of various calibres; nucleated blood cells; presence of a fibrous capsule formed by 3 to 4 layers of collagen fibres interspersed with spindle cells and non-vacuolized hepatocytes.

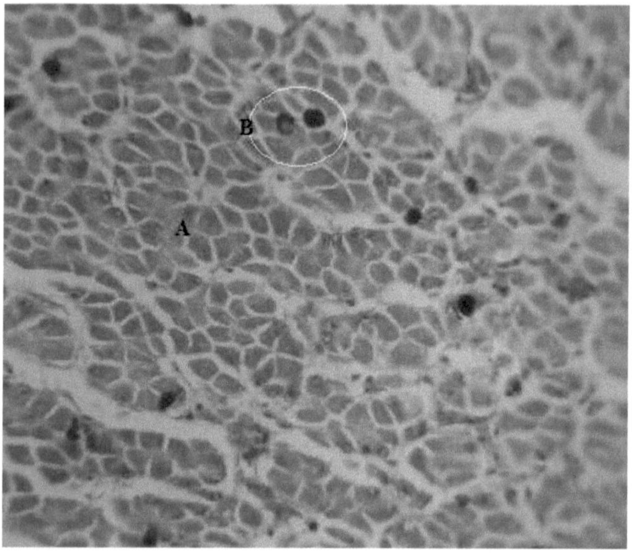

Figure 29: Photomicrograph of the histological section of the liver of *L. latrans,* stained with HE under a 4x objective. A - non-vacuolized hepatocytes; B - presence of granulations forming densely stained, larger and more condensed extracellular corpuscles. Source: Nascimento and Lima, 2010.

Figure 29 was taken to illustrate a histological section of the liver of animals with similar descriptions, which were sampled on slides numbers 24, 79 and 96.

The environment where these animals were captured was characterized by the presence of sparse riparian vegetation, an area of pasture, an agricultural plantation approximately 500 metres away and irrigation channels with six pumps located at the end of the Urubu River and the beginning of the Formoso River, Fazenda Formosa, Lagoa da Confusao - TO.

The animals belonged to point 3, an area where pesticides were present. The rostral-cloacal length and the month of capture were, respectively: 9.6 cm - April; 6.2 cm - July; 8.4 cm - October, all of which are considered to be medium-sized individuals, since these anurans can reach up to 15 cm in their adult phase. The Lepdactylidae is made up of tiny (12 mm) to huge (25 cm) anurans, from all habitats and with different modes of reproduction (POUGH *et al.*, 2003).

Among all the histological liver sections analyzed, there were some animals with minimal alterations, which had the structure described in table 5, line 1, found in animals (24, 29, 79 and 96), which represents a standard closer to normal for a *L. latrans,* i.e. free from major environmental stresses and aggressions. Even so, these animals showed alterations, not least because they belonged to a region where pesticides are used.

Liver cells are responsible for a number of functions that are essential for the survival of the species. They play an important role in the metabolism of proteins, lipids and carbohydrates, serve as storage sites for many nutrients, such as glucose, and detoxify the body. In addition, they are involved in hematopoiesis (the process of forming, maturing and releasing blood cells into the bloodstream) during larval life and in the secretion of bile (KUMAR; ANSARI, 1986).

Not all pesticides cause the immediate death of fish and sublethal doses can cause other disturbances to the health of these animals, so histological changes in fish tissues are a sensitive tool for detecting the direct toxic effects of chemical compounds on target organs and are powerful indicators of previous exposure to environmental stressors (HINTON *et al.*, 1992; SCHWAIGER, *et al.*, 1997).

Hepatocytes can be considered primary targets of toxic agents, which makes them suitable as biomarkers of environmental pollution (BRAUNBECK, 1998).

Even though this is considered to be normal, it was possible to find alterations in these animals. The presence of granulations forming densely stained and sparsely dispersed extracellular corpuscles was seen (figure 29). This alteration usually occurs when the liver is unable to expel the rest of the residues of metabolism, which can be characterized as some toxic residue. The granulation is therefore nothing more than residues of metabolism that have not been eliminated. The liver alterations caused by glyphosate/Roundup®, although reversible, compromise the normal functioning of the organ (ALBINATI *et al.*, 2007).

Consequently, the granulations may have been the result of exposure to toxins from pesticides used in the agricultural plantations located at the point where these animals were captured, since they were caught in areas of the irrigated rice, soybean and bean project in the irrigation canals. The fact that no major alterations were found may be due to the length of time they were exposed, as they are medium-sized animals.

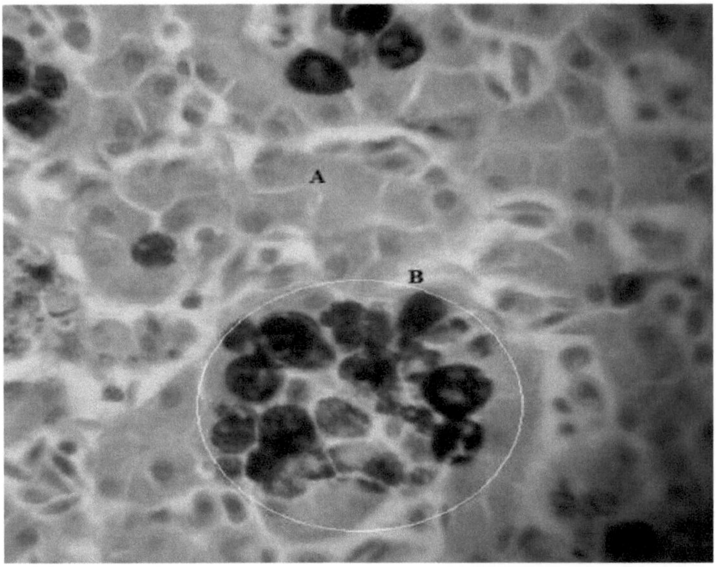

Figure 30: Photomicrograph of the histological section of the liver of *L. latrans,* stained with HE under a 40x objective. A - non-

vacuolized hepatocytes; B - presence of granulations forming densely stained, larger and more condensed extracellular corpuscles. Source: Nascimento and Lima, 2010.

Figure 30 shows *L. latrans* animals caught at different points that have the same histological characteristics:

Animal 23 - point 1 - rostro-cloacal - 9.1 cm - March

Animal 29 - point 2 - rostro-cloacal - 8.6 cm - March

Animal 77 - point 3 - rostro-cloacal - 8.0 cm - July

Animal 78 - point 3 - rostro-cloacal - 7.9 cm - July

Based on the above information, it is possible to say that the animals are of a size that characterizes them as medium-sized, captured in an area with pesticides, referring to points 2 and 3 and one animal from point 1, a place with no pesticides.

Animal 23, the largest of them, even though it was captured at a pesticide-free point, showed significant alterations in the liver, which may be related to the presence of clay extraction activity, where machinery used in the region to remove the material was constantly found. At the site, it was noticed that some products were used, such as motor oil, because the packaging was abandoned at the site after extraction. Another hypothesis could be that the animals had migrated from areas where they were suffering some kind of exposure to various toxins. This hypothesis is less likely to have occurred, since the site was 17 km from the urban area and the nearest farms were those that only worked with cattle.

The histological characteristics described are similar to those of figure 29, previously analyzed, where the following were also identified: Presence of classic hepatic lobules and hepatic portal triad. Polyhedral hepatocytes organized in acini, identifiable space of Disse, filled with sinusoids. Hepatic sinusoids passing through the hepatocytes, presence of arteries of various sizes, presence of hepatic veins of various sizes, nucleated blood cells, presence of a fibrous capsule formed by 3 to 4 layers of collagenous fibers interspersed with spindle cells. Hepatocytes are not vacuolized.

The alteration found was the same as in figure 29, however the presence of granulations forming the extracellular corpuscles was striking due to the fact that they were densely stained, larger in size and more condensed.

One of the animals analyzed here refers to a spot that is supposed to be free of pesticides, as it is not close to an agricultural plantation, and which even showed large quantities of granulations.

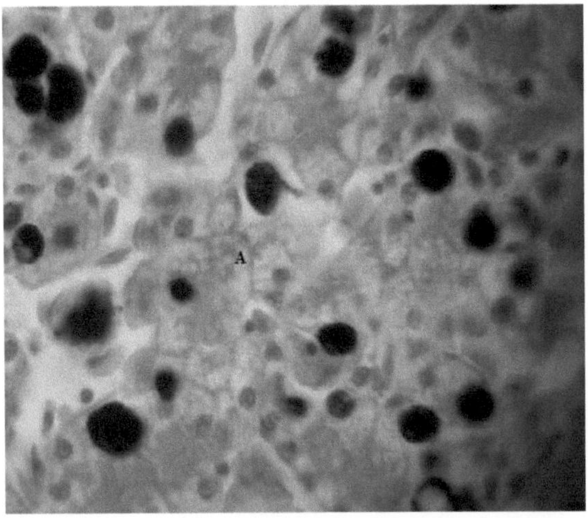

Figure 31: Photomicrograph of the histological section of the liver of *L. latrans,* stained with HE under a 4x objective. A compartmentalization; B granulations, C - relative cytoplasmic vacuolization of the

Figure 31 refers mainly to animals from points 2 and 3. However, it also shows an animal from point 1, an area supposedly free of pesticides, which was also captured in March and measured 8.4 cm, considered to be young. The other animals were all from points 2 and 3 and varied in size, caught in May, July and September, as detailed below:

Animal 22 - point 1 - 8.4 cm - March

Animal 46 - point 2 - 3.4 cm - May

Animal 60 - point 2 - 9.1 cm - July

Animal 67 - point 2 - 8.0 cm - July

Animal 86 - point 3 - 6.2 cm - July

Animal 92 - point 2 - 3.2 cm - September

Animal 101 - point 3 - 5.0 cm - September

In the histological section of the liver shown in figure 29, it was possible to see the presence of compartmentalization of the hepatic lobules, the presence of a hepatic portal triad; with evidence of alterations such as: the presence of granulations forming densely stained extracellular corpuscles; compartmentalization; relative cytoplasmic vacuolization of the hepatocytes.

Vacuolization is the blank space around the nucleus of the hepatocyte, which appears in the form of bubbles, characteristic of cell death. This is a sign of alteration/aggression, which can be caused by longer exposure to toxins.

Compartmentalization of the hepatic lobules is also a change in the level of hepatocytes, which is the formation of compartments (groups) of hepatocytes.

The regenerated liver is usually well organized, exhibiting the typical lobular arrangement and consequently normalized function. However, when hepatocytes are repeatedly attacked over a long period, their multiplication is followed by a significant increase in the amount of connective tissue. Instead of the normal organization of the liver lobules, nodules of different sizes form, many of which are visible to the naked eye. These nodules are composed of a central mass of hepatocytes in disorganized arrangements, surrounded by a large amount of dense connective tissue. It is a diffuse fibrosis, affecting the entire liver, resulting from various conditions that affect liver architecture (JUNQUEIRA E CARNEIRO, 2008).

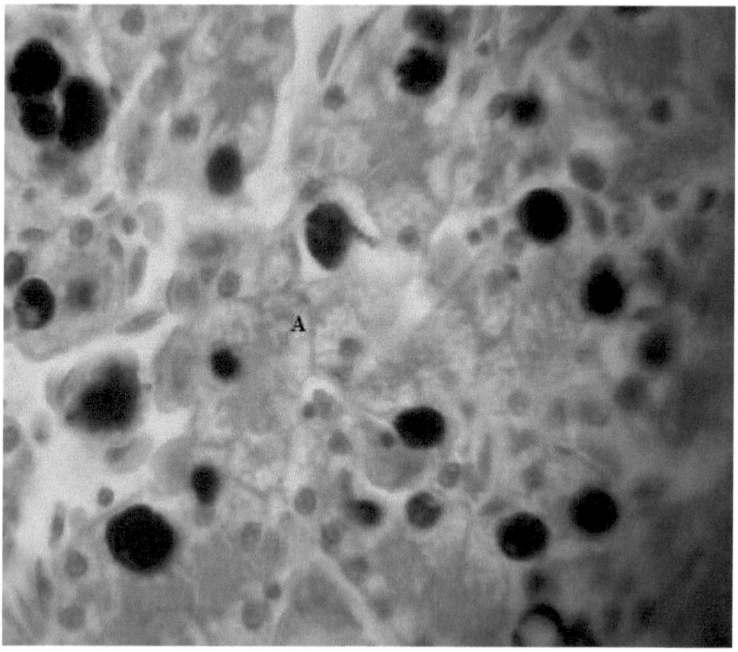

Figure 32: Photomicrograph of the histological section of the liver of *L. latrans,* stained with HE under a 40x objective. A - Intense vacuolization of the hepatocytes. Source: Nascimento and Lima, 2010.

Figure 32 shows the histological section of the liver of animals from Point 1, an area supposedly free of pesticides, captured in March and October, with measurements ranging from 8.3 to 9.5 cm.

Animal 20 - point 1 - 8.3 cm - March

Animal 94 - point 1 - 8.4 cm - October

Animal 95 - point 1 - 9.5 cm - October

The histological characteristics of the liver of these animals differ from those of previous animals, which showed the hepatic portal triad, i.e. a structure made up of a vein, duct and artery.

Despite being animals representative of pesticide-free areas, histological analysis showed that the animals had alterations at the level of the liver, such as: compartmentalization of the hepatic lobules; a large amount of granulations forming densely stained extracellular corpuscles; intense vacuolization of the hepatocytes.

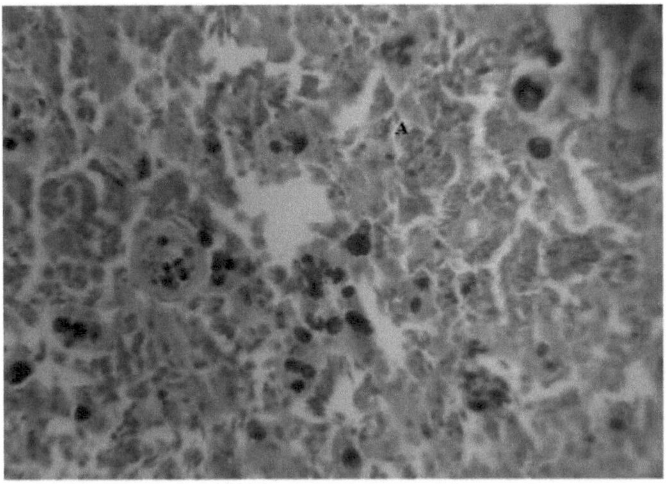

Figure 33: Photomicrograph of the histological section of the liver of *L. latrans,* stained with HE under a 10x objective. A - Intense hyperemia with loss of the vascular framework. Source: Nascimento and Lima, 2010.

Only one of the animals analyzed had a hepatic system structure, as shown in figure 33. Animal 26 was a sample from point 2, captured in the embankment used for irrigation of the flood-irrigated rice agricultural project, in March, and measured 9.5 cm.

Histologically, the liver section showed compartmentalization of the hepatic lobules, intense hyperemia and loss of the vascular framework; presence of granulations forming densely stained extracellular corpuscles with relatively medium to large sizes. These alterations are considered critical, leading us to believe that the time and doses of toxins to which the animal was exposed were high.

Hyperemia is the loss of the vascular framework, i.e. blood spilling into the Tisse space.

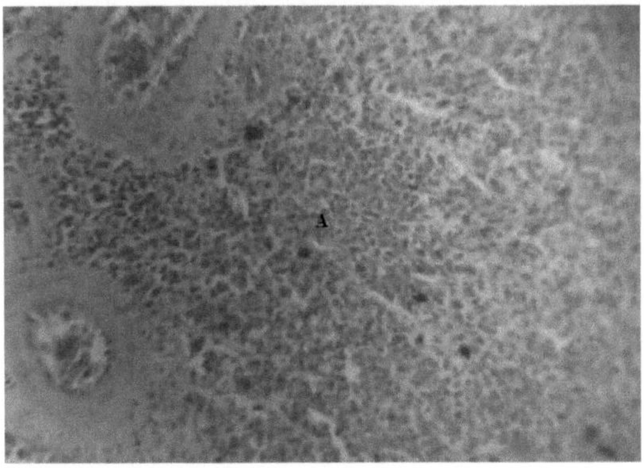

Figure 34: Photomicrograph of the histological section of the liver of *L. latrans,* stained with HE under a 4x objective. A Leukocyte infiltrate. leukocyte infiltrate. Source: Nascimento and Lima, 2010.

Figure 34 shows the livers of animals from points 2 and 3, the area with the presence of an agricultural project, which varied in size and showed the same characteristics in both young and adult animals in the months of April, May, July and October, as shown below:

Animal 3 - Point 3 - rostro-cloacal - 5.1 cm - April

Animal 6 - Point 3 - rostro-cloacal - 5.9 cm - April

Animal 11 - Point 3 - rostro-cloacal - 7.9 cm - May

Animal 49 - Point 2 - rostro-cloacal - 3.5 cm - May

Animal 61 - Point 2 - rostro-cloacal - 9.0 cm - July

Animal 66 - Point 2 - rostro-cloacal - 9.0 cm - July

Animal 81 - Point 3 - rostro-cloacal - 6.1 cm - July

Animal 82 - Point 3 - rostro-cloacal - 7.6 cm - July

Animal 97 - Point 3 - rostro-cloacal - 9.4 cm - October

Significant changes were identified in these animals, such as: compartmentalization of the hepatic lobules, relative hyperemia with loss of the vascular framework and leukocyte infiltrate; presence of granulations forming densely stained extracellular corpuscles with relatively medium to large sizes. The changes are even more significant than those found in animals from point 1, as shown in figure 34.

In general, nocturnal species spend the day in crevices or buried in places where there is a suitable microclimate (high humidity, moderate temperature, lack of sunlight). At dusk, when the dangers of drying out and the increased humidity of the environment disappear, amphibians become active, as long as the environmental conditions associated with the season allow (JORGE, 1986).

Methyl parathion causes high percentages of inhibition in the activities of cholinesterases and amnotransferases (*B. cephalus*), a tropical fish native to the Amazon region. Hematological responses have shown that high concentrations of the pesticide can cause damage to hematopoietic tissue (AGUIAR *et al.*, 2000).

The discrete vacuolization found in the liver of animals exposed to pesticides suggests a mobilization of glycogen due to chemical stress (CONSTANTINO, 2007). With this in mind, it is clear from this study that the animals were suffering from the pesticides used in the irrigation projects, as well as other chemical compounds present at the three points studied.

Therefore, in relation to the liver, the tissues of the animals that were analyzed for the three points studied showed alterations in their composition: the presence of granulations forming the densely stained and poorly dispersed extracellular corpuscles, as well as more condensed and of varying sizes; compartmentalization of the hepatic lobules; intense vacuolization of the hepatocytes; relative and intense hyperemia with loss of the vascular framework and leukocyte infiltrate.

One study revealed that the liver accumulates high concentrations of toxins, which is possibly related to its role as the main storage and regulatory organ in compound homeostasis, reflecting a multifunctional action in detoxification processes (CARPENÈ and VASÂK, 1989). In the liver of *L. latrans, it* was possible to see alterations in the liver, which were probably caused by pesticides from the agricultural project, as well as by oil used in the clay extraction area.

In general, the absorbed compound is transported via the bloodstream to the liver where it is metabolized and excreted via the bile. Accumulation occurs in tissues when the liver's maximum storage capacity is reached (LAURÈN AND MCDONALD, 1987). No signs of alterations were found in the female or male reproductive system.

These alterations, depending on the length of exposure to stress and the amount of toxins received, can lead to organ damage, significantly impairing the performance of the animal's physiological activities. Depending on the pesticide to which they are exposed, they can produce histological alterations that lead to tumors. Another consequence of the presence of contaminants in these tissues is that they are liable to be passed on in the food chain to predators of *L. latrans* (GRISOLIA, 2005).

Skin

Anurans absorb liquids through their skin, which is why they are always near streams. When they are far from rivers, they absorb water from their own urine, which is stored in their bladder. Their skin is permeable, which means that substances such as oxygen, water and even dangerous chemicals can be absorbed through it. Their body has ectothermy, a variation in temperature according to the ambient temperature. When winter comes, many hibernate, i.e. take a long sleep, where the heart beats slowly and breathing is also very slow, so the body temperature drops, and during this period the body feeds on stored fat, until the weather gets warm again. Amphibians, classified as ectotherms, get their heat mainly from external sources, either from the sun or from resting on a heated rock (POUGH *et al.*, 2003).

The presence of such permeable skin in amphibians poses problems that make it difficult for them to survive on land: the loss of large amounts of water through perspiration can lead to dehydration, which explains their dependence on water and their preference for humid environments (SILVA & SASSON, 2005). The evaporation rate of anurans is of the same order of magnitude as evaporation on a surface of free water and their skin does not seem to represent a significant barrier to evaporation (POUGH *et al.*, 2003).

Chart 6: Histological description of the skin of the animals analyzed, Lagoa da Confusao and Cristalândia, 2009.

No. Figure	Cuts (làmina)	Histological pattern observed	Possible changes
Figure 35	3; 20; 22; 26; 46; 61; 82; 94;	One layer of cells in the stratum corneum, two layers of cells in the intermediate stratum with a flat nucleus, basal stratum (normal), large glands near the basal lamina, melanin around the glands.	Melaninano around glands.
Figure 36	23; 92;	One keratinized layer, three layers of the intermediate stratum with flattened nuclei, basal stratum with heterochromatic ovoid nuclei (normal), absence of vacuolization (normal); absence of melanin near the basal lamina.	
Figure 37	6; 28; 29; 49; 95;	One keratinized layer; cells of the intermediate stratum with ovoid nuclei; cells of the basal stratum with large, heterochromatic spherical nuclei; medium cytoplasmic vacuolization; sub-epithelial glands; accumulation of melanin near the basal lamina.	Basal stratum cells with large, heterochromatic spherical nuclei; medium cytoplasmic vacuolization;

			accumulation of melanin near the basal lamina.
Figure 38	11; 24; 66; 67; 77; 78; 79; 81; 86; 96; 97; 101;	Absence of keratinized layer, basal layer with large or bulky euchromatic nuclei, large cytoplasmic vacuolization, with large subepithelial glands, abundance of melanin near the basal lamina.	large cytoplasmic vacuolization; abundance of melanin near the basal lamina

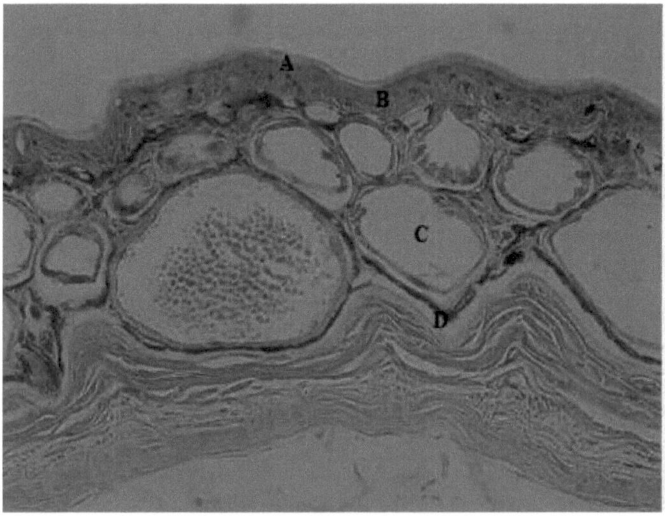

Figure 35: Photomicrograph of the histological section of the skin of *L. latrans,* stained with HE under a 4x objective. A. Layer of cells in the stratum corneum; B. Layers of cells in the intermediate stratum with flat nuclei; C. Large glands near the basal lamina; D. Melanin around the glands. Source: Nascimento and Lima, 2010.

Figure 35 shows animals caught at the three study points during the months of March, April, May, July and October 2009, classified as juveniles and adults according to their measurements, as can be seen below:

Animal 3 - Point 3 - rostro-cloacal - 5.1 cm - April

Animal 20 - point 1 - rostro-cloacal - 8.3 cm - March

Animal 22 - point 1 - rostro-cloacal - 8.4 cm - March

Animal 26 - point 2 - rostro-cloacal - 9.5 cm - March

Animal 46 - point 2 - rostro-cloacal - 3.4 cm - May

Animal 61 - point 2 - rostro-cloacal - 9.0 cm - July

Animal 82 - point 3 - rostro-cloacal - 7.6 cm - July

Animal 94 - point 1 - rostro-cloacal - 8.4 cm - October

Histological sections of the skin of these animals showed the following structure: one layer of cells in the stratum corneum, two layers of cells in the intermediate stratum with a flat nucleus, basal stratum (normal), large glands near the basal lamina, melanin around the glands.

There were histopathological alterations in the skin of these animals, such as the presence of melanin around the glands, which means that these animals experienced some stress or aggression in their environment, which triggered a defense reaction that they have - the production of melanin.

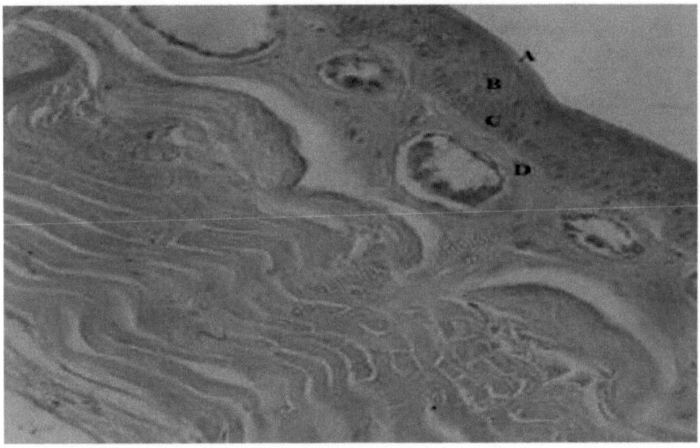

Figure 36: Photomicrograph of the histological section of *L. latrans* skin stained with HE under a 10x objective. A. Keratinized layer; B. Intermediate stratum layers with flattened nuclei; C. Basal layer with heterochromatic ovoid nuclei; D. Absence of melanin near the basal lamina. Source: Nascimento and Lima, 2010.

The animals were captured at two different points, one in an agricultural project area and the other in an area of native vegetation, but with the presence of clay extraction, one young and the other adult, during the months of March and September (figure 36).

Animal 23 - point 1 - rostro-cloacal - 9.1 cm - March

Animal 92 - point 2 - rostro-cloacal - 3.2 cm - September

Their skin structure was as follows: one keratinized layer, three layers of the intermediate stratum with flattened nuclei, basal stratum with heterochromatic ovoid nuclei (normal), absence of vacuolization (normal); absence of melanin near the basal lamina. In this case there was no evidence of histological changes.

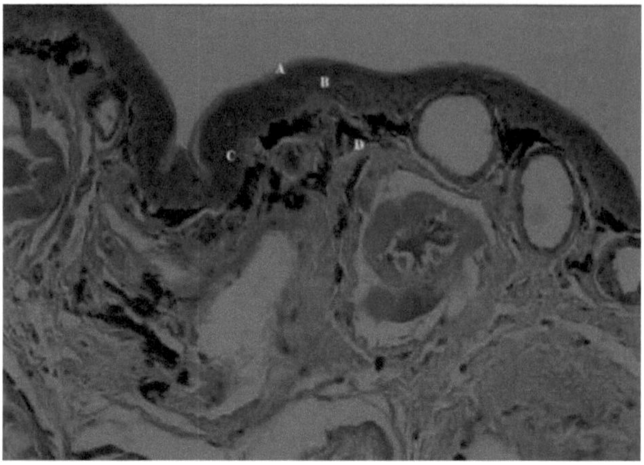

Figure 37: Photomicrograph of the histological section of *L. iatrans* skin stained with HE under a 4x objective. A - A keratinized layer; B - Cells of the intermediate stratum with ovoid nuclei; C - Cells of the basal stratum with large, heterochromatic spherical nuclei; D - Accumulation of melanin near the basal lamina. Source: Nascimento and Lima, 2010.

The animals with this histological skin pattern were caught in March, April, May and October 2009, at points 1, 2 and 3, some of which were young (Figure 37).

Animal 6 - Point 3 - rostro-cloacal - 5.9 cm - April

Animal 28 - point 2 - rostro-cloacal - 8.0 cm - March

Animal 29 - point 2 - rostro-cloacal - 8.6 cm - March

Animal 49 - Point 2 - rostro-cloacal - 3.5 cm - May

Animal 95 - point 1 - rostro-cloacal - 9.5 cm - October

The histological characteristics presented were: a keratinized layer; cells of the intermediate stratum with ovoid nuclei; cells of the basal stratum with large, heterochromatic spherical nuclei; medium cytoplasmic vacuolization; sub-epithelial glands; accumulation of melanin near the basal lamina.

The histopathological changes found in these animals were: basal stratum cells with large, heterochromatic spherical nuclei; medium cytoplasmic vacuolization; melanin accumulation near the basal lamina, which may be a possible reaction triggered by the environmental stress experienced.

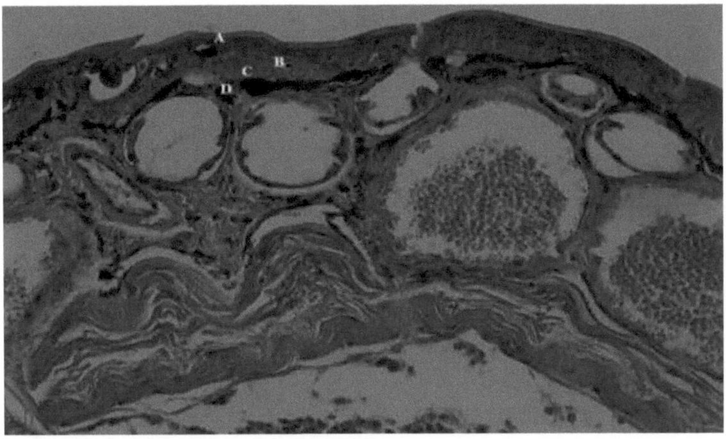

Figure 38 Photomicrograph of the histological section of *L. latrans* skin stained with HE under a 4x objective. A - Absence of keratinized layer; B - Large cytoplasmic vacuolization; C - Basal layer with large, bulky euchromatic nuclei; D - Abundance of melanin near the basal lamina. Source: Nascimento and Lima, 2010.

The animals that showed the histological pattern represented by figure 38 were only from points 2 and 3, where there was an agricultural project area, captured in the months of May, April, July and October 2009.

Animal 11 - Point 3 - rostro-cloacal - 7.9 cm - May

Animal 24 - point 3 - rostro-cloacal - 9.6 cm - April

Animal 66 - Point 2 - rostro-cloacal - 9.0 cm - July

Animal 67 - point 2 - rostro-cloacal - 8.0 cm - July

Animal 77 - point 3 - rostro-cloacal - 8.0 cm - July

Animal 78 - point 3 - rostro-cloacal - 7.9 cm - July

Animal 79 - point 3 - rostro-cloacal - 6.2 cm - july

Animal 81 - Point 3 - rostro-cloacal - 6.1 cm - July

Animal 86 - point 3 - rostro-cloacal - 6.2 cm - July

Animal 96 - point 3 - rostro-cloacal - 8.4 cm - October

Animal 97 - Point 3 - rostro-cloacal - 9.4 cm - October

Animal 101 - point 3 - rostro-cloacal - 5.0 cm - September

Histologically, the animals above showed: absence of a keratinized layer, basal layer with large or bulky euchromatic nuclei, large cytoplasmic vacuolization, with large subepithelial glands, abundant melanin near the basal lamina.

In general, the histopathological alterations found in the skin slides analyzed were: abundance of melanin near the basal lamina; absence of a keratinized layer; basal layer with large or bulky euchromatic nuclei.

In this sense, in addition to the acute toxicity that aquatic animals can face due to the presence of chemical products in the water of the Urubu River and irrigation channels, they can suffer chronic effects when exposed to low concentrations or sub-lethal concentrations of toxic agents over a long period of time.

Some highly toxic chemical compounds have a harmful intra- and extracellular action on aquatic organisms, which can cause the mucus in tadpoles to coagulate on the animal's skin, impairing oxygenation, since in tadpoles there is thermal exchange, which represents 30% of total respiration (STORER and USINGER, 1979; LIMA and AGOSTINHO, 1989). The presence of melanin is an animal's defense mechanism against the stresses it experiences. As we can see, an accumulation of melanin was found in the skin of the animals at the three points, leading us to believe that the chemical compounds present in these locations may have led to this reaction. These changes were noticed not only in the young animals, but also in the adults analyzed at points two and three, although they were only noticeable in the adult animals at point 1 (an area of native vegetation and with the presence of clay extraction activities), which may suggest that the stress did not affect the young animals analyzed at this point.

Reproductive Systems

There are two types of reproduction in amphibians, explosive and prolonged, with very short reproduction lasting only a few days and very long reproduction lasting several months, respectively (POUGH *et al.*, 2003).

The reproduction of some species occurs after atmospheric precipitation, in temporary marshes, or in permanent marshes, which provide a greater supply of water (SCHIMIDT, 1999).

Fertilization is external, the males, in copula, pour their seminal fluid onto a gelatinous cord that surrounds the eggs as they exit the female's cloaca. Once fertilized, the eggs receive different care, depending on the species, where they develop in the vocal sacs, in recesses of the dorsal skin, wrapped around the legs or simply tangled in aquatic plants, thus forming the so-called tadpoles, remaining in the water until metamorphosis (SILVA & SASSON, 2005).

Chart 7: Histological description of the female reproductive system of the animals analyzed, Lagoa da Confusao and Cristalândia, 2009.

Figure number below	Cuts (blade)	Histological pattern observed	Possible changes
Figure 39	46; 67;	Uterine tube with simple cylindrical epithelium with steriocilia, goblet cells, oocytes with decentralized and heterochromatic nucleus, chromophobic cytoplasm.	Normal
Figure 40	20; 78; 96;	Ovocytes in various stages surrounded by a membrane of flat cells and some blood capillaries; unidentified ovary, unidentified fallopian tubes.	Normal

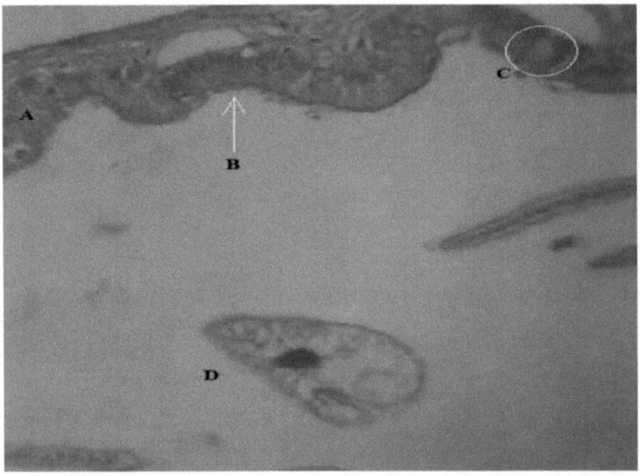

Figure 39: Photomicrograph of the histological section of the reproductive system of a female *L. latrans* stained with HE under a 40x objective. A - uterine tube with simple cylindrical epithelium; B - steriocilia; C - goblet cells; D - oocytes with decentralized nucleus, chromophobic cytoplasm. Source: Nascimento and Lima, 2010.

Figure 39 shows the reproductive system of females found at Point 2, an agricultural plantation area, in May and July, one of which was a juvenile and the other an adult.

Animal 46 - point 2 - rostro-cloacal - 3.4 cm - May

Animal 67 - point 2 - rostro-cloacal - 8.0 cm - July

In the slides with histological sections of the female reproductive system, it was possible to identify the presence of a uterine tube with a simple cylindrical epithelium with steriocilia, goblet cells, oocytes with a decentralized and heterochromatic nucleus, chromophobic cytoplasm, all with a normal appearance.

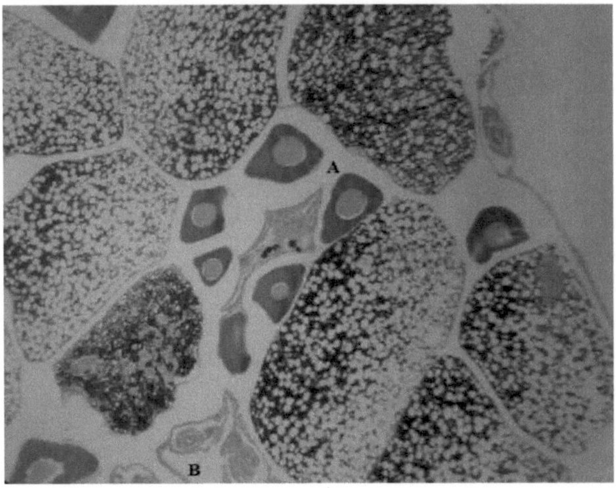

Figure 40: Photomicrograph of the histological section of the reproductive system of a female *L. latrans* stained with HE under a 10x objective. A - Ovocytes in various stages; Membrane of flat cells and some blood capillaries.

Source: Nascimento and Lima, 2010.

The animals found at points 1 and 3, an area of native vegetation and an agricultural project, respectively, were adult females, captured in March, July and October 2009 (Figure 40).

Animal 20 - point 1 - rostro-cloacal - 8.3 cm - March

Animal 78 - point 3 - rostro-cloacal - 7.9 cm - July

Animal 96 - point 3 - rostro-cloacal - 8.4 cm - October

Histologically, the females showed oocytes, in various stages, surrounded by a membrane of flat cells and some blood capillaries. The ovaries and fallopian tubes were not identified in the histological slides of the animals in question.

Chart 8: Histological description of the male reproductive system of the animals analyzed, Lagoa da Confusao and Cristalândia, 2009.

Figure number below	**Cuts (làmina)**	**Histological pattern observed**	**Possible changes**
Figure 41	24; 28; 97;	Testicles subdivided into lobes, presence of albuginea tunic formed by collagen fibers and cells with flattened nuclei interspersed with the fibers. Accumulation of spermatogonia near the basal lamina; presence of spindle cells around the seminiferous tubule; presence of nerve ganglia in the testicular mediastinum; presence of small *Leydig* cells in the testicular interstitium; Each lobe is filled with a large number of seminiferous tubules in which there are Sertoli cells supporting a large number of germ cells differentiated into flagellated cells (spermatids) highly grouped on the Sertoli cells and also in the lumen of the tubules, indicating an advanced stage of spermatogenesis.	Normal
Figure 42	26; 95; 101;	Testicles subdivided into lobes. Accumulation of spermatogonia near the basal lamina; presence of spindle cells around the seminiferous tubule; presence of nerve ganglia in the testicular mediastinum; presence of small *Leydig* cells in the testicular interstitium; presence of albuginea tunic formed by collagen fibers and cells with flattened nuclei interspersed with the fibers. Each lobe is filled with a large number of seminiferous tubules in which *Sertoli* cells appear internally, supporting a large number of germ cells in different processes of cell differentiation, presenting spherical cells and also flagellated cells, indicating an intermediate stage of spermatogenesis.	Normal
Figure 43	11; 29; 49; 61; 79	Testicles subdivided into lobes; accumulation of spermatogonia near the basal lamina; presence of spindle cells around the seminiferous tubule; presence of nerve ganglia in the testicular	Normal

		mediastinum; presence of small *Leydig* cells in the testicular interstitium; presence of an albuginea tunic formed by collagen fibers and cells with flattened nuclei interspersed with the fibers; each lobe is filled with a large number of seminiferous tubules in which *Sertoli cells* appear internally, supporting a large number of spherical germ cells that have not yet differentiated into spermatids, demonstrating the initial stage of spermatogenesis.	
Figure 44	22; 81, 94;	Immature testicle. Thick tunica albuginea, seminiferous tubules separated by a large number of collagen fibers and blood vessels of various calibers; internal lining formed by basophilic cuboidal cells and absence of spermatogenesis.	Normal

Testicle

The basic organization of the testicle is common to all amphibians. It has spermatogenic and androgenic functions and two main compartments: the interstitial and the tubular. The interstitial or intertubular compartment contains blood vessels, nerve fibers, connective tissue cells and fibers, as well as Leyding cells, which have a steroidogenic function. The production of androgens is important for sexual differentiation, the development of secondary sexual characteristics and sexual behavior and for the regulation of spermatogenesis (WELTZIEN et al., 2004).

The tubular compartment contains somatic cells (Sertoli cells) and the germ cells that will form the spermatozoa after undergoing a very complex and highly organized process, spermatogenesis (BILLARD, 1990; KOULISH, et al., 2002).

Sertoli cells physically and functionally delimit a clone of germ cells at the same stage of development that originates from a single primary spermatogonia, thus forming the spermatozoa or spermatogenic cyst. Thus, Sertoli cells provide germ cells with physical support and important factors for survival, proliferation and differentiation, and are also involved in hormonal intermediation and phagocytosis of cell debris originating from spermiogenesis (WELTZIEN *et al.*, 2004).

The Sertoli cells that delimit the spematogenous cysts are supported by their own tunic, which is made up of the basal membrane (acellular layer) and myoid peritubular cells that have contractile capacity (LE GAC and LOIR, 1999).

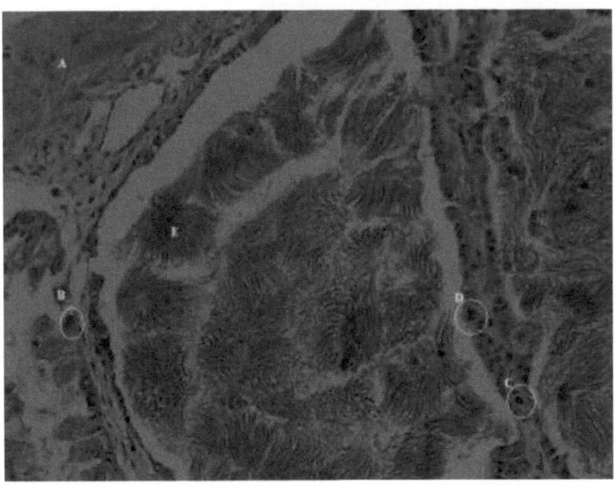

Figure 41: Photomicrograph of the histological section of the immature testicle of *L. iatrans* stained with HE under a 10x objective. A - Tunic albuginea formed by collagen fibers; B - Spermatogonia near the basal lamina; C - Presence of small *Leydig cells* in the testicular interstitium; D - *Sertoli cells* supporting a large number of germ cells; E - Spermatids highly grouped on the *Sertoli cells*. Source: Nascimento and Lima, 2010.

The animals captured at points 2 and 3, in April, March and October, at the points with agricultural plantations, were adult males (Figure 41).

The fact that amphibians are ectothermic animals (whose temperature is usually slightly above the ambient temperature) probably means that the duration of spermatogenic events varies according to temperature, being faster at higher temperatures (FRANÇA *et al.*, 2002).

Animal 24 - point 3 - rostro-cloacal - 9.6 cm . April

Animal 28 - point 2 - rostro-cloacal - 8.0 cm - March

Animal 97 - Point 3 - rostro-cloacal - 9.4 cm - October

The reproductive system of these males had testicles subdivided into lobes, presence of albuginea tunic formed by collagen fibers and cells with flattened nuclei interspersed with the fibers; accumulation of spermatogonia near the basal lamina; presence of spindle cells around the seminiferous tubule; presence of nerve ganglia in the testicular mediastinum; presence of small *Leydig* cells in the testicular interstitium; each lobe is filled with a large number of seminiferous tubules in which *Sertoli cells* appear supporting a large number of germ cells differentiated into flagellated cells (spermatids) highly grouped on the Sertoli cells and also in the lumen of the tubules indicating an advanced stage of spermatogenesis.

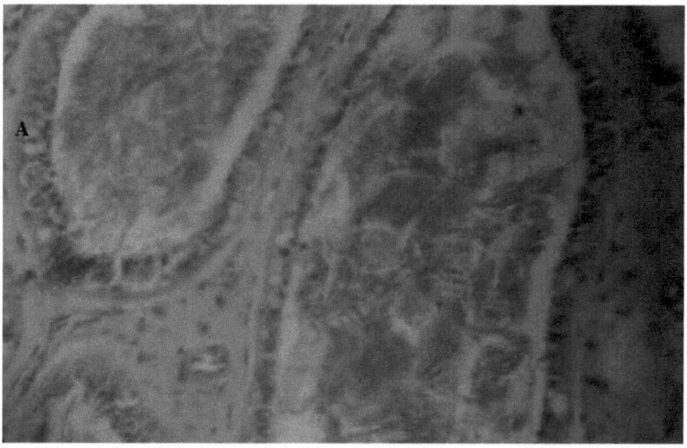

Figure 42: Photomicrograph of the histological section of the immature testicle of *L. latrans,* stained with HE under a 10x objective. A - *Sertoli* cells supporting a large number of germ cells in different cell differentiation processes, indicating an intermediate stage of spermatogenesis. Source: Nascimento and Lima, 2010.

The animals shown in figure 42 are young and adult males from points 1, 2 and 3 in March, September and October 2009.

Animal 26 - point 2 - rostro-cloacal - 9.5 cm - March

Animal 95 - point 1 - rostro-cloacal - 9.5 cm - October

Animal 101 - point 3 - 5.0 cm - September

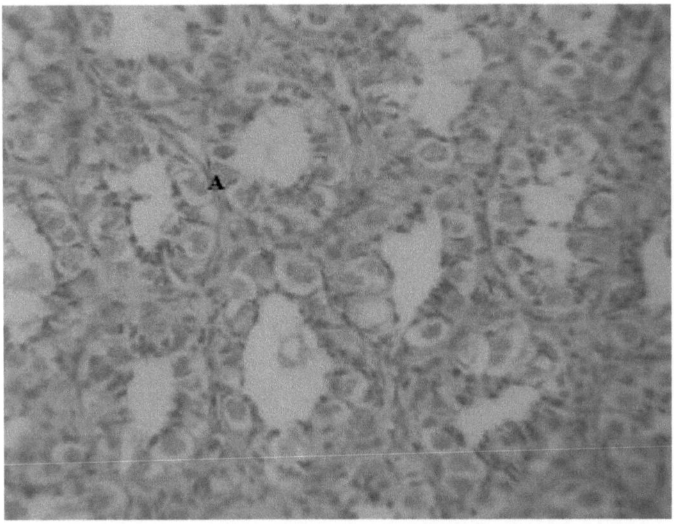

Figure 43: - Photomicrograph of the histological section of the immature testicle of *L. latrans* stained with HE under a 10x objective. A - Sertoli cells supporting a large number of spherical germ cells that have not yet differentiated into spermatids, showing an early stage of spermatogenesis. Source: Nascimento and Lima, 2010.

In the reproductive system of these animals, it was possible to identify testicles subdivided into lobes. Accumulation of spermatogonia near the basal lamina; presence of spindle cells around the seminiferous tubule; presence of nerve ganglia in the testicular mediastinum; presence of small *Leydig* cells in the testicular interstitium; presence of albuginea tunic formed by collagen fibers and cells with flattened nuclei interspersed with the fibers. Each lobe is filled with a large number of seminiferous tubules in which *Sertoli* cells appear internally, supporting a large number of germ cells in different processes of cell differentiation, presenting spherical cells and also flagellated cells, indicating an intermediate stage of spermatogenesis.

Figure 43 shows the reproductive system found in young and adult males at points 2 and 3 in March, May and July 2009.

Animal 11 - Point 3 - rostro-cloacal - 7.9 cm - May

Animal 29 - point 2 - rostro-cloacal - 8.6 cm - March

Animal 49 - Point 2 - rostro-cloacal - 3.5 cm - May

Animal 61 - point 2 - rostro-cloacal - 9.0 cm - July

Animal 79 - point 3 - rostro-cloacal - 6.2 cm - july

Histologically, they had Testicles subdivided into lobes. Accumulation of spermatogonia near the basal lamina; presence of spindle cells around the seminiferous tubule; presence of nerve ganglia in the testicular mediastinum; presence of small *Leydig* cells in the testicular interstitium; presence of albuginea tunic formed by collagen fibers and cells with flattened nuclei interspersed with the fibers. Each lobe is filled with a large number of seminiferous tubules in which *Sertoli* cells appear internally, supporting a large number of spherical germ cells which have not yet differentiated into spermatozoa, demonstrating the initial stage of spermatogenesis.

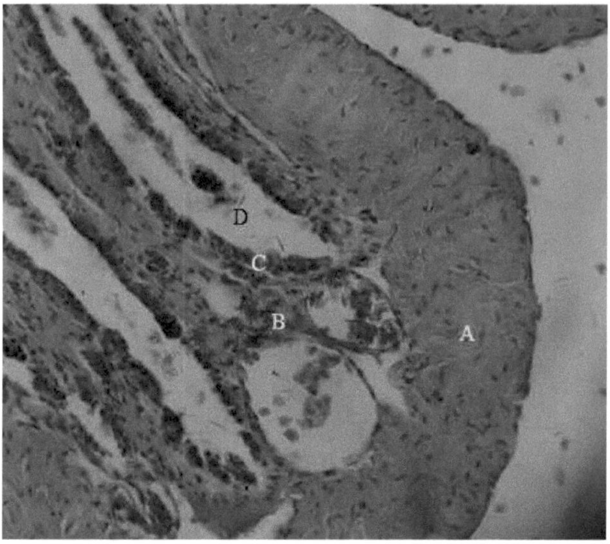

Figure 44 Photomicrograph of the histological section of the immature testicle of *L. latrans,* stained with HE under a 40x objective. A. Thick tunica albuginea; B. Seminiferous tubules separated by a large amount of collagen fibers; C. Internal lining formed by basophilic cuboidal cells; D. Absence of spermatogenesis. Source: Nascimento and Lima, 2010.

The animals shown in Figure 44 were males (young and adult), caught at points 1 and 3 in March, July and October 2009.

Animal 22 - point 1 - rostro-cloacal - 8.4 cm - March

Animal 81 - Point 3 - rostro-cloacal - 6.1 cm - July

Animal 94 - point 1 - rostro-cloacal - 8.4 cm - October

It has an immature testicle; a thick tunic albuginea, seminiferous tubules separated by a large number of collagen fibers and blood vessels of various calibers; an internal lining formed by basophilic cuboid cells and the absence of spermatogenesis.

The main factors that may be causing the decline in amphibian populations are: acid rain due to the release of nitrogen and sulphur oxides from burning fossil fuels, which alters the pH of water in rivers and lakes, affecting the development of eggs; increased ultraviolet radiation, due to the destruction of the ozone layer, which also affects egg development, and deforestation, which causes a reduction in vegetation cover, changing the temperature and relative humidity of the air, eliminating the specific microenvironments of amphibians (POUGH *et al.*, 1999 cited by FERES, 2001).

One of the factors that significantly alters amphibian behavior is temperature (SOARES & BATISTA, 2004), a consequence of changes in the forest canopy that generate hot, dry conditions for amphibians, due to the increased incidence of ultraviolet rays (POUGH *et al.*, 2003).

Among the discharges carried out by agricultural activity, the greatest proportions are pesticides and inputs, which have negatively affected all ecosystems, terrestrial, aquatic and aerial, making aquatic pollution one of the most worrying

(COSTA, 2004).

The increase in temperature influences the increase in toxicity of organisms in aquatic environments. If a fish living in an environment that receives warm water has to devote more of its metabolic energy to compensating for the stressful effects of the chemical compound and the high temperature, it will consequently not have enough energy for the activities necessary for survival. Adaptation becomes increasingly costly as the limiting conditions, such as temperature, approach extremes and the organism becomes increasingly susceptible to other factors such as disease, predation and toxins (ODUM, 1987).

Aquatic biota is constantly exposed to toxic substances released into the environment from various emission sources, such as heavy metals, pesticides, organic compounds, among others, which contribute to the contamination of aquatic ecosystems. These substances are capable of interacting with living organisms, causing multiple alterations that can have serious consequences for populations, communities or ecosystems, depending on the degree of contamination and the length of exposure (ARIAS, 2002).

Around 70% of the water in Brazil is used by agriculture, and the majority of this is used for irrigation, and comes from the conservation of rivers, which are constantly being degraded by the deforestation of riparian forests and the dumping of pollutants on a scale greater than the maximum absorption limit of the rivers, altering their dynamics. The pesticide residue can bind to suspended particulate matter, settle in the sediment or be absorbed by organisms, and can then be accumulated, transported through the aquatic system by diffusion in water currents or in the bodies of aquatic organisms. Some pesticides and/or metabolites can also return to the atmosphere through volatilization (COSTA, 2004).

In this way, it is possible to identify the interaction that exists between pesticides, sediment and water, influenced by water movement, turbulence and temperature (WHITE & RASMUSSEN, 1985). This interaction can result in a longer exposure time for aquatic organisms to toxic compounds, where pesticides can reach aquatic environments through intentional application, drift and surface runoff from areas where applications have taken place (ZAMBRONE, 2009).

4 - CONCLUSION

Among the tissues analyzed, the liver and skin of *L. latrans* showed histopathological alterations, which were more critical in the liver of the animals for the three points analyzed.

Points 2 and 3, which are areas with agricultural plantations, are probably attributed to the use of pesticides. For point 1, which is an area free of pesticides, as it is a remote region with a considerable degree of conservation, it is attributed to the existence of an activity that was carried out in times of drought, clay extraction, and the substance that could be causing the alterations are oils used in the machinery, which was noticed by the disposal made at the site after the end of the activities.

The male and female reproductive systems showed a normal histological pattern.

The changes caused are directly related to toxicity, exposure time and the speed of penetration through cell envelopes. The likelihood that adverse ecological effects may occur as a result of the exposure of natural ecosystems to one or more stressors can cause severe risks to other biological communities and even to human health.

BIBLIOGRAPHICAL REFERENCES

AGUIAR, L. H. DE.; CORRÊA, C. F.; MORAES, G. Effects of the organophosphate pesticide Methyl Parathion (Folidol 600[R]) on the metabolism and cholinesterase activity of the Freshwater Teleost, Brycon cephalus (matrinxa)

(Gunther, 1869). Federal University of Sao Carlos. In: **Ecotoxicology**: perspectives for the 21st century. 2000.

AICHINGER, M. 1987. **Annual activity patterns of anurans in a seasonal neotropical environment**. Oecologia. 71: 583-592.

ALBINATI, A.C.L.; MOREIRA, E.L.T.; ALBINATI, R.C.B. et al. Acute toxicity of Roundup® herbicide to piauçu (*Leporinus macrocephalus*). **Revista Brasileira de Saùde** Prod. Anim., v.8, 2007, p.184-192.

ARIAS, A.R.L.; VIANA, T.A.P.; INACIO, A.F. **Utilization of Bioindicators as tools for environmental monitoring and evaluation**: the case of water resources. 2002. Available at: < http://www.ebape.fgv.br/radma/doc/FET/FET-020.pdf>. Accessed on: July 8, 2008.

AURICCIHIO P.; SALOMÂO M. da G. **Procedure for collecting samples of *L. latrans*:** Vertebrate collection and preparation techniques. 2002.

BILLARD, R. Spermatogenesis in teleost fish. In: LAMMING G.E. (ed). **Reproduction in the male**. Churchill Livingstone, 1990.

BRAUNBECK,T. Cytological alterations in fish hepatocytes following *in vivo* and *in vitro* sublethal exposure to xenobiotics - structural biomarkers of environmental contamination. In: BRAUNBECK, T.; HINTON, D.E.; STREIT, B. **Fish ecotoxicology**. Berlin: Birkhauser, 1998. p. 61-140.

BURKETT, D. W. and THOMPSON, B. C. 1994. **Wildlife association with human-altered water sources in semiarid vegetation communities**. Conserv. Biol. 8: 682-690.

CARPENE, M.; VASAK, M. Hepatic Metallothionin from Goldfish (*Carassius auratus*). **Comp. Biochem. Physiol.**, 92B, 1989. 463-468.

CONSTANTINO, L. S. **Evaluation of oxidative stress in *Geophagus brasiliensis*, exposed to pesticides in irrigated rice cultivation, in the municipality of Ararangù, SC**. University of the Extreme South of Santa Catarina. Santa Catarina. 2007

COSTA, L.M.; GJORUP, G.B. Environmental problems caused by agriculture in the Cerrado. In: AGRICULTURE AND THE ENVIRONMENT, 1994, Viçosa. **Anais**...Viçosa: Nùcleo de Etudos e Pesquisa em Meio Ambiente - NEPEMA/UFV, 2004. p. 101-112.

DIXO, M.B.O. **Effect of forest fragmentation on the community of frogs and burrowing lizards in southern Bahia**. Master's dissertation, USP. 77p. 2001.

FANTA, E.; RIOS, F. S.; ROMÂO, S.; VIANNA, A. C. C.; FREIBERGER, S. Histopathology of the fish *Corydoras paleatus* contaminated with sublethal levels of organophosphrus in water and contaminated with sublethal levels of organophosphorus in water and food. **Ecotoxicology and Environmental Safety**. V54. 2003.

FEIO, R. N.; BRAGA, U. M. L.; WIEDERHECKER, H. & Santos, P. S. **Anfibios do Parque Estadual do Rio Doce** (Minas Gerais). Federal University of Viçosa, State Forestry Institute, MG, 1998. 32p.

FILONOV, A.G., Tereshchencko, I.E., Monzón, C.O., Gonzâlez, R.M.E. and Godinez, D.E. 2000. Seasonal variability of temperature and salinity fields in the coastal zone of the states of Jalisco and Colima, Mexico. **Ciências Marinas**. 26(2):303-321.

FRANÇA, L. R.; ROCHA, D. C. M.; MIRANDA, J. R.; DEBELJUK, L. Sertoli cell proliferation and testicular function. **Boletin Informativo de la Sociedad Argentina de Andrologia**. V11. 2002.

GASCON, C. 1991. **Population- and community-level cloacalyses of species occurrences of Central Amazonian rainforest tadpoles**. Ecology 72: 1731-1746.

GEORGE, L. L. e CASTRO, R. R. L. **Histologia comparada.** 2. Ed. rev. e ampl. Sao Paulo: Roca, 1998.

GENESER, F. **Histology**. On biomolecular bases. 3. ed. Ed. Médica Panamericana. Buenos Aires. Argentina. 1999.

HEALH, A. G. **Water pollution and fish physiology**. Boca Raton: CRC Press, 1987.

HEYER, W. R. 1973. Ecological interactions of frog larvae at a seasonal tropical location in Thailand. J. Herpetol. 7: 337-361.

HINTON,D.E.; BAUMANN,P.C.; GARDNER,G.R.; HAWKINS,W.E.; HENDRICKS,J.D; MURCHELANO,R.A. ; OKIRINO,M.S. Histopathology biomarkers. In: HUGGET,R.J.; KIMERLE,R.A.; MEHRLE Jr,P.M.; BERGAMAN, H. L. **Biomarkers biochemical, physiological and histological markers of anthroponenic stress**. Florida: Lewis Publishers,. 1992. p.155-209.

JIM, J. 1980. **Ecological aspects of amphibians recorded in the region of Botucatu, São Paulo (Amphibia, Anura).** Doctoral dissertation, University of São Paulo, Sâo Paulo.

JORGE, J. L. T., **Manual para el censo de los vertebrados terrestres,** Editora Raices, Madrid, 1986, p. 38.

JUNQUEIRA, L. C.; CARNEIRO, J. **Histologia bàsica**. 11. ed. Rio de Janeiro: Guanabara Koogan, 2008.

JUNQUEIRA, L. C. U.; CARNEIRO, J. Histologia bàsica. 11 ed. Rio de Janeiro: Guanabara Koogan, 2008.

KOULISH, S.; KRAMER C. R. GRIER, H. J. Oragnization of the male gonad in a protogynous fish, *Thalassoma bifasciatum* (Teleostei: Labridae). **Journal of Morphology**, v.254, 2002.

KRUGER, R. A. **Analysis of the toxicity and genotoxicity of pesticides used in agriculture using bioassays with *Allium cepa***, 2009.

KUMAR, K.; ANSARI, B. A. Malathion toxicity: effect on the liver of the fish Brachydanoi rerio (Cyprinidae). **Ecotoxicology and environmental Safety.** Orlando, v.12. 1986.

LAURÈN, D. J., MCDONALD, D. G. Acclimation to copper by rainbow trout, ***Salmo gairdneri:*** physiology. Can. J. Fish. Aquat. Sci., 42: 630-648.

LIMA, S.L.; AGOSTINHO, C.A. **A** Criaçâo **de Ras**, 2°ed., Sâo Paulo: Globo, 187p., 1989.

LE GAC, F.; LOIR, M. Male reproductive system, fish. In: KNOBIL, NEILL, J. D. (ed.).

Encyclopedia of reproduction. San Diego: Academic Press. v.3 p.20-30. 1999.

LOMBARDI, J.V. **Fundamentals of Aquatic Toxicology.** In: RANZANI-PAIVA,M.J.T.; TAKEMOTO, R.M. and LIZAMA, M.A.P. Sanidade de Organismos Aquàticos, Editora Varela, p.263272, 2004.

MARZON. A. F., PINHEIRO, G.H.D., FERNANDES, M. N. Contamination of aquatic ecosystems through the study of copper toxicity in Curimbatà, *P. Scrofa* (Teleostei, Prochilodontidae). Federal University of São Carlos. In.: In.: **Ecotoxicology**: perspectives for the 21st century. 2000.

MILLER, G. T.. Environmental Science. Technical review: Thomson Learning, 1931. Original Title: **Environmental science** "Translated from the II[a] norie-american edition". 2007.

ODUM, E. P. Ecology. Editora Guanabara, Rio de Janeiro - RJ. p. 434.

POUGH, F. H.; HEISER, J. B.; MCFARLAND, W. N. **The Life of Vertebrates**. 2nd Edition. Atheneu Editora, Sâo Paulo, 1999. 798 p.

POUGH, F.H.; JANIS, C.M.; MESER, J.B. **A vida dos vertebrados**. Sâo Paulo: Atheneu, 2003.

RABB, G. B. **Declining amphibian population. Species**, 1990. 33-34.

SANTOS, E., **Anfibios e Répteis**, Editora Villa Rica, Volume 3, 4ª Edition, 1994, 263 p.

SCHIMIDT, N. K., **Fisiologia Animal - Adaptaçao e Meio Ambiente**, Editora Santos, Sâo Paulo, 1999, 600 p.

SCHWAIGER, J.; WANKE, R.; ADAM, S.; PAWERT, M.; HONNEN, W.; TRIEBSKORN, R.; The use histopathological indicators to evaluate contaminant related in fish. **Journal of Aquatic Ecosystem Stress and Recovery**, v.6, p. 75-86, 1997.

SILVA Jùnior, C. & Sasson, S., **Biologia 2,** Editora Saraiva, 8ª Ediçâo, 6ª tiragem Sâo Paulo, 2005, 527 p.

SOARES, R. V. & Batista, A. C., **Meteorologia e Climatologia Florestal,** Universidade Federal do Paranà, 2004, 195 p.

STEBBINS, R. C.; COHEN, N. W., **A Natural History of Amphibians**. Princeton University Press. New Jersey, 1995, 316p.

STORER, T.I.; USINGER, R.L. **Zoologia Geral**. Sâo Paulo, Companhia Editora Nacional, 1979. 757p.

TOCANTINS. Planning and Environment Secretariat. **Diagnosis of the use of agrochemicals in the sub-basins upstream of the Cantao State Park and its surroundings.** Palmas: SEPLAN, 2002.

US Environmental Protection Agency (USEPA). **Proposed guidelines for ecological risk assessment**: Notice. FRL-5605-9. *Federal Register*, 61, 47552-47631.1996.

US Environmental Protection Agency (USEPA). Proposed guidelines for ecological risk assessment: Notice. FRL-5605-9. **Federal Register**, 61, 47552-47631. 1996.

VOLNEI, W. G.; SIQUEIRA, W. C. Basic histotechnology. Laboratory of the pathological anatomy and cytology unit of the Hospital de Base do Distrito Federal. Brasilia - DF. 2000.

WATSON, G.F.; DAVIES, M.; TYLER, M. J. 1995. **Observations on temporary waters in northwestern Australia. Hydrobiologia.** 299: 53-73.

WELTZIEN, F. A.; ANDERSSON, E.; SHALCHIAN-TABRIZI, K.; NORBERG, B. The brain pituitary gonad axis in male teleosts, with special emphasis on flatfish (Pleuronectiformes). **Compatative Biochemistry and Physiology** - part A: Molecular & Integrative Physiology, v.137, n.3 p. 447-477. 2004.

WHITE, P. A; RASMUSSEN, J. B. The Genotoxic Wastes in Surface Waters. **Mutat Res.**, Madrid, v. 410, p. 223-236, Jan. 1998.

WOEH Jr, G.; WOEH, E. N. Caracteristicas dos amfibios. Available at http://www.ra- bugio.org.br/anfibios sobre.php. Accessed on March 2, 2008.

YAHKSON, K.; MOYSE, J. Cryopreservation of the spermatozoa of Crassotrea tulipa and three other oysters. **Aquaculture**, 97: 259-267. 1991.

ZAMBRONE, F. A. D. SINTOX: **Toxicological Information System**. Rio de Janeiro: Fiocruz, 2002, p.2-43. Available at: <http://www.fiocruz.br.cict.oque.estrut.dect/sintox. kitbrasil.html > Accessed on January 2, 2009.

ZIMMERMAN, B. L. and RODRIGUES, M. T. Frogs, snakes, and lizards of the INPA - WWF Reserves near Manaus, Brazil. In: Gentry, A.H. (ed.), Four Neotropical

Rainforests. Yale University Press, New Raven. 1990.

ANNEXES

ANNEX I: RAN/SISBIO/IBAMA license

IBAMA - Online Services

CPF: **866.442.381-49**

Name: **Raquel Aparecida Mendes Lima**

Ibama registration: **2843442**

Online Services Manual
Change Password
Exit

Date accessed: **15/05/2009 21:07:19**

Authorizations and licenses for scientific and educational purposes - SISBIO

Path- Services >> Authorizations and licenses for scientific and educational purposes - SISBIO >> Select : request >> Authorization for activities for scientific purposes >> Request extract

Request extract

Basic data

Request no.: **16736**

Type of request: **Authorization for scientific activities**

Situation: **Document granted**

Title: **Evaluation of water contamination by pesticides, correlated to the presence of Leptodactylus latrans, in the Urubu river sub-basin - TO.**

The request involves endangered species: **NO**

Institution: **FUNDACAO UNIVERSIDADE FEDERAL DO TOCANTINS**

Researcher details

Researcher: **Raquel Aparecida Mendes Lima**

CPF: **866.442.381-49**

Address: **103 Norte Rua no 03 Lote 29**

Neighborhood: **Plano Diretor Norte**

City: **PALMAS- TO**

ZIP CODE: 77001-018

Phone: (0XX63) 9998-2908

FAX: (0XX63) 3218-2198

E-mail: raquelengenheira@gmail.com

PLACE(S) OF RESEARCH

N°	Place of research	Biome	Municipality	UF	Situation
1	State of Tocantins	Cerrado	LAGOA DA CONFUSAO	TO	Authorized
2	State of Tocantins	Cerrado	CRISTALANDIA	TO	Authorized
3	State of Tocantins	Cerrado	FATIMA	TO	Authorized

TAXON(S) X ACTIVITIES INCLUDED IN THE REQUEST

N°	Tàxon	Top rating	Planned quantity	Activity	Situation
1	Species Leptodactylus latrans	Amphibians	---	Capturing wild animals in situ	Authorized
2	Species Leptodactylus latrans	Amphibians	---	Collection/transport of biological samples in situ	Authorized
3	Species Leptodactylus latrans	Amphibians	100	Collection/transport of wildlife specimens in situ	Authorized
4	Species Leptodactylus latrans	Amphibians	---	Tagging wild animals in situ	Authorized

TAXONS X MATERIAL, METHODS, BIOLOGICAL SAMPLES OF THE REQUEST

N°	Taxonomic group	Description	Type	Situation
1	Amphibians	Animal dead orparts (carcass)/bone/skin	Biological samples	Authorized
2	Amphibians	Biopolymer	Marking method	Authorized
3	Amphibians	Manual capture	Method capture/collection	Authorized
4	Amphibians	Puça	Method capture/collection	Authorized

DESTINATION(S) OF BIOLOGICAL MATERIAL(S)

N °	Destination	Type of destination	Situation

1	FEDERAL UNIVERSITY OF TOCANTINS	collection	Authorized
2	university of sao paulo	USP Histology Laboratory	Initial state

PARTICIPATING INSTITUTION(S)

N°	Name of institution	Participation	Situation
1	UFG	Perform histological analysis	Authorized

SCHEDULE OF ACTIVITIES

N °	Activity description	Start date	End Date	Situation
1	Bibliographic review	01/03/2008	31/12/2009	Authorized
2	Demarcation of anuran and water collection points	05/01/2009	09/01/2009	Authorized
3	Collection of water samples at delimited points on the Urubu River	02/02/2009	31/10/2009	Authorized
4	Campaigns to capture and morphologically analyze L Leptodactyloides and ocellatus	02/02/2009	31/10/2009	Authorized
5	Analysis of physico-chemical and biological parameters of water samples collected	02/02/2009	31/10/2009	Authorized
6	Chromatography	02/02/2009	31/10/2009	Authorized
7	Histological analysis of L. leptodactyloides and ocellatus samples	01/10/2009	31/10/2009	Authorized
8	Tabulation of data and submission of dissertation	03/10/2009	28/03/2010	Authorized

SUBJECTS OF THE REQUEST

N°	Subject description	Situation
1	Conservation	Authorized
2	Ecology	Authorized
3	Herpetology	Authorized
4	Hydrology	Authorized
5	Histology	Authorized
6	Inventory	Authorized

7	Morphology	Authorized
8	Toxicology and Exotic Plants	Authorized
9	Zoology	Authorized

CHAPTER IV GENERAL DISCUSSION

Man depends on the soil and the soil depends on man and his use of it. Even if pesticides are used properly, they end up causing problems. When pesticides are applied to crops, what doesn't volatilize and run off into waterways ends up in the soil, initiating a series of processes that can lead to contamination and influence the biological balance in the soil (MAIRESSE *et al.*, 2007).

The Urubu River has several large agricultural projects, which vary according to the time of year, defined by the amount of water available. During the rainy season they usually grow flood-irrigated rice and during the dry season they work with soybeans, beans, corn and watermelons.

The river rises in the municipality of Fàtima do Tocantins (Figure 1 - delimitation of the study area in Chapter I) and the irrigation pumping sites are present from the stretches that comprise the municipality of Lagoa da Confusao - TO. As a result, the use of pesticides in that region is constant and unavoidable, as we witnessed during the interviews and from the accounts of those interviewed for this research.

The strategy used to verify the contamination of the water was to identify retailers, producers and histopathologically analyze common animals in the region, which successfully achieved the objective. From the studies carried out, it was possible to see histological changes in tissues which, according to the literature, are actually affected by the presence of pesticides in water.

Pesticide poisoning can occur during handling, transportation and application, especially if they are highly toxic. Mild poisoning of the mother's body can result in serious damage to the fetus or death. Chronic damage results from prolonged ingestion of persistent poisons in the environment and/or food, as they accumulate in the body, in tissues and organs (ALMEIDA, 1982).

According to the interviews with farmers, the points where pesticides were found most often were 2 and 3, which the visits showed are the points where irrigation projects are being carried out. These were also the points where the most animals were found with histopathologies in the liver, as well as in the skin.

The mechanisms of action of herbicides on plants can also be indicative of the degree of toxicity to other living beings and aggressiveness to the environment (MAIRESSE *et al.*, 2007). *L. latrans* was considered to be a significant animal for this study, due to its dependence on the aquatic environment for activities necessary for its survival, such as reproduction.

The dependence of amphibians on the aquatic and terrestrial environment is a strong environmental indicator, since any disturbance caused in these environments can directly affect the balance of these animals (DIXO, 2001). The deposition of pesticides and oil derivatives can quickly decimate entire populations of amphibians, even if the quantities of these products eliminated are negligible. Their reproductive and feeding behaviors contribute directly to the reduction of populations, as they do not adapt to changes in their natural environments (BARBOSA *et al.*, 2007).

In this sense, and in addition to knowing that there is a possibility that the water pumped from the Urubu River into the irrigation canals will return in some form, and the possibility of finding the animals in these places, the chance of finding alterations in these animals would increase, as in fact occurred.

The numbers of *L. latrans* recorded for the agricultural project areas are considerably high, since the intervention provided by the agriculture present in the area is a limiting factor for the development of amphibian populations. Another factor that makes them more sensitive to environmental variations is their dependence on the aquatic environment, and they are easily eliminated when they come into contact with pesticides. Despite all the anthropogenic intervention that has taken place on the site through the monoculture of rice for six months, and soybeans, beans, corn or watermelons in the second half of the year, *L. latrans* were easily observed on the margins of the plantations. Unlike the site with the presence of native vegetation, the encounter of these animals was rarer.

Depending on the contaminant, it can remain in the animal's tissues for a longer period of time, offering a carcinogenic risk, for example (TAYLOR, 1978), which may be responsible for the alterations found in the liver and skin. Due to the high bioaccumulative power of pesticides, animals are more susceptible to possible poisoning, generating multiple serious alterations with consequences for populations, communities or ecosystems, depending on the degree and time of exposure (ARIAS, 2002).

Histological changes in the tissues of organisms dependent on aquatic environments are a sensitive tool for detecting the direct toxic effects of chemical compounds on target organs, and are powerful indicators of previous exposure to environmental stressors (HINTON *et al*., 1992; SCHWAIGER, *et al.,* 1997).

The stress of these animals indicates that it may have been caused by a chemical group present in the study areas due to the histopathological alterations found such as: mild to strong vacuolization of the liver, which is treated as a strategic tissue for identification because it is an organ that functions as a filter in the body.

In the light of the research carried out, it was possible to verify that some form of contamination of environmental compartments is occurring, including the aquatic one, since alterations in the environmental quality indicator used in this study were proven.

However, in order to diagnose with greater precision the origin of the alterations found in the liver and skin, it would be necessary to carry out more in-depth studies, such as chromatography on both animals (*L. Iatrans*) and water.

CHAPTER V GENERAL CONCLUSION

The amphibian population serves to balance the ecosystem by controlling insects and other invertebrates, as well as being a menu in the food chain for reptiles, birds and mammals. Without amphibians, the Yavouras would be subject to pest infestations and humans would not survive the many diseases transmitted by insects. The alteration of rivers to form lakes or lagoons, deforestation, the use of agrochemicals and pollution have all been target problems for the extermination of amphibians. If some areas without degradation are cloacalyzed, we can infer that the biomass of amphibians will be greater than that of other animals (ra-bugio.org.br, 2007).

The interview process shows that it is of fundamental importance to demonstrate the need for environmental education practices with all those involved in these projects, which will certainly take some time for them to become deeply involved for generations and generations, until something more positive prevails in the lives of those people.

BIBLIOGRAPHICAL REFERENCES

ALMEIDA, W. F. de. Toxicity of pesticides, their control and restriction of sale. In. GRAZIARO NETO, F (Coord.). **Uso de agrotóxico e receituàrio agronòmico.** Sao Paulo: Agroediçoes, 1982. p.59-73.

ARIAS, A.R.L.; VIANA, T.A.P.; INÂCIO, A.F. **Utilização de Bioindicadores como ferramentas de monitoramento e avaliação ambiental:** o caso de recursos hydricos. 2002. Available at: < http://www.ebape.fgv.br/radma/doc/FET/FET-020.pdf>. Accessed on: July 8, 2008.

BARBOSA, H. V. M., RABELLO, H., SAMPAIO, F. D. F., CASTRO, T. M., MAIOLLI, L. U., PEREIRA, E. A. Survey of the anurofauna of the legal reserve of Fazenda brunoro agro-avicola in venda nova do Imiogrante, State of Espirito Santo. **Anais.** VIII Congress of Ecology of Brazil. Caxambu - MG. 2007.

DIXO, M. B. O. **Effect of forest fragmentation on the community of leaf litter frogs and lizards in southern Bahia**. Master's dissertation - University of Sao Paulo. Sao Paulo. 2001.

HINTON,D.E.; BAUMANN,P.C.; GARDNER,G.R.; HAWKINS,W.E.; HENDRICKS,J.D; MURCHELANO,R.A. ; OKIRINO,M.S. Histopathology biomarkers. In: HUGGET,R.J.; KIMERLE,R.A.; MEHRLE Jr, P.M.; BERGAMAN, H. L. Biomarkers biochemical, physiological and histological markers of anthroponenic stress. Florida: Lewis Publishers,. 1992. p.155-209.

MAIRESSE, L. A. S.; COSTA, E. C.; FARIAS, J. R.; FIORIN. R. A. Bioactivity **of plant extracts on lettuce (Lactuca *sativa* l.**1) **Bioactivity of plant extracts on lettuce (Lactuca *Sativa l.*).** Revista da FZVA. Uruguaiana, v.14, n.2, p. 1-12. 2007.

SCHWAIGER, J.; WANKE, R.; ADAM, S.; PAWERT, M.; HONNEN, W.; TRIEBSKORN, R. The use of histopathological indicators to evaluate contaminant-related stress in fish. J. Aquat. Ecosyst Stress Recov, v. 6, n. 1, p. 75-86, 1997.

TAYLOR, G. R. **A ameaça ecològica.** ed. Brasileira rev. prefacada por Màrio G. Ferri; traduçao de Rodrigo Machado. Sao Paulo: Verbo: Ed. da Universidade de Sao Paulo, 1978.

MIX
Papier aus verantwortungsvollen Quellen
Paper from responsible sources
FSC® C105338

Printed by Books on Demand GmbH, Norderstedt / Germany